High-Level Synthesis

Blue Book

Michael Fingeroff

Library of Congress Control Number: 2010906504
ISBN: Hardcover 978-1-4500-9724-6
Softcover 978-1-4500-9723-9
E-book 978-1-4500-9725-3

This book was printed in the United States of America.

To order additional copies of this book, contact:
Xlibris Corporation
1-888-795-4274
www.Xlibris.com
Orders@Xlibris.com
80137

CONTENTS

"You can't always get what you want
But if you try sometimes you just might find
You get what you need "

The Rolling Stones

Preface

As a former RTL designer who made the switch to High-Level Synthesis (HLS) years ago, I can still remember both the excitement of discovering a totally new design methodology, as well as the frustration of not knowing what to expect from the HLS tool as I attempted to code increasingly complex designs.

My first exposure to the world of HLS began with a demo of a FIR filter written in C++, and synthesized directly to RTL. This was a design I was very familiar with coming from the world of RTL. After having written numerous hand-coded VHDL and Verilog implementations, I was completely blown away when I saw how a single algorithmic C++ description could be used to generate a large number of RTL implementations, each with different area/performance characteristics. The HLS tool was able to do in a matter of minutes, what normally would take me days or weeks to accomplish. The fact that it was then able to take the resulting RTL and simulate it using my C++ testbench to prove functionality was simply unbelievable. The potential of the technology was too attractive for me not to make the switch, and I soon found myself working in the field of HLS.

As I began to tackle more complex designs, I started to encounter occasional problems with achieving the best possible results. What I saw was that the style in which my C++ was written could have a big impact on the resulting quality of the RTL. I would liken this to what I had experienced when people were first making the switch from schematic based design to RTL, where proper coding style was critical to good quality of results. Fortunately for me, I was working with a number of experts in HLS who I could rely on to provide the explanation as to why a particular coding style gave less than desirable results, and who could suggest a better way to write the C++. I was lucky since at that time there was no formal style guide for writing C++ for synthesis.

As HLS has matured the quality of results has improved dramatically for a much wider range of C++ coding styles. However, this does not mean that all styles are equal, and there is still the potential for ending up with poor quality RTL when the C++ is not well written. Good style not only requires an understanding of the underlying hardware architecture of an algorithm, so that it is reflected in the C++ design, but also an understanding of how HLS works.

This book presents the recommended coding style for C++ synthesis that results in good quality RTL. Most of the C++ examples are accompanied with hardware and timing diagrams, where appropriate. The basic concepts of HLS are introduced and an effort is made to relate them directly to concepts that are well understood by RTL engineers. Although this book focuses primarily on C and C++ to illustrate the fundamentals of C++ synthesis, all concepts presented here are equally applicable to SystemC when describing the core algorithmic part of a design. Although the examples are simplistic in many cases, they illustrate the fundamental principles behind C++ hardware design. These concepts will translate to much larger designs.

As a final thought for the RTL Designers, System Architects, and Algorithm Designers who are looking to adopt High-Level Synthesis; correct-by-construction RTL synthesized from C++ may not look exactly like what you would code by hand. HLS optimizations often can result in the odd logic gate in the resulting schematic. The hardware diagrams in this book exclude all of the extraneous logic. You will probably end up saying at some point that “It’s not exactly what I expected the RTL to look like”. However you should also ask yourself “How long would it have taken to write using hand-coded RTL?”, “Would it have simulated correctly the first time?” and ultimately “Is it good enough?”. If you don’t get too caught up in the details you’ll find the results to be exactly what you need.

Mike Fingeroff, January 2010

Who Should Read This Book

Engineering managers should read Chapter One to understand how HLS evolved from existing design methodologies and how it can help improve current design flows. RTL designers should read the entire book and System/Algorithm designers should read chapters 3, 4, 5, 8, and 9, at the very least.

About the Authors

Mike Fingeroff has worked as a technical marketing engineer for the Catapult C product line at Mentor Graphics since 2002. His areas of interest includes DSP and high-performance video hardware. Prior to working for Mentor Graphics he worked as a hardware design engineer developing real-time broadband video systems. Mike Fingeroff received both his bachelors and masters degrees in electrical engineering from Temple University in 1990 and 1995 respectively.

Thomas Bollaert is product marketing manager for the Catapult C product line at Mentor Graphics. He has a more than 15 years of experience in EDA, and an extensive background in system-level design and high-level synthesis. More recently, Thomas worked in tight collaboration with Mentor Graphics' European customers, helping them learn, adopt and deploy high-level synthesis to improve their design practices. He earned his electronic engineering degree from ESIEE Paris where he specialized in hardware architectures for signal processing applications.

Acknowledgements

I would like to thank all of the people that contributed to the creation of this book by taking their valuable time to review the content and provided essential feedback.

Emmanuel Liegeon, Thales Alenia Space
Mathieu Lebon, Alyotech
Katsunobu Natori, Hitachi Ltd.

Special thanks to Shawn McCloud, the High-Level Synthesis Product Line Director at Mentor Graphics, for providing me with the time and resources to write this book.

I would also like to thank all of my co-workers at Mentor Graphics for providing internal review of the book: Suravinth Sundralingam, David Burnette, Tony Vandinh, Mike Bradley, Mike Hilsen, and Bob Condon.

Lastly I would like to thank Thomas Bollaert for his contribution of the first chapter of this book as well as his input on the cover artwork. Thanks to Lucien Murray-Pitts for both his review and contribution to the technical content. Special thanks to Bryan Bowyer and Peter Gutberlet for answering numerous questions about HLS and coding style. Thanks to Ron Plyler for his discussion on pipeline feedback, and finally thanks to Andres Takach, Stuart Clubb, and Tom Nagler for their contributions to the chapter on FFT transforms.

Chapter 1

Making the Case for High-Level Synthesis

The promise of high-level synthesis (HLS) is a powerful one: the ability to generate production-quality register transfer level (RTL) implementations from high-level specifications. In other words, HLS automates an otherwise manual process, eliminating the source of many design errors and accelerating a very long and iterative part of the development cycle.

A broken design flow

To fully understand the potential and benefits of HLS it is important to put things in the perspective of a hardware design flow. Today, most projects start with some form of specification. Sometimes this is a simple, written document, but quite frequently an executable model is created—usually in ANSI C, C++ or SystemC. At this early stage, the specification is essentially functional: it contains little to no hardware implementation details, and its primary purpose is to validate and fine-tune the desired behavior. Once tested, this behavioral model undergoes a several step process until it takes the form of the actual hardware implementation. The first step is to define an optimal architecture to implement the desired functionality. If the functionality defines "what" the system does, the architecture defines "how" the system does it, with direct consequences on performance, area, and power consumption. After the architecture is defined, the design team hand-codes these decisions in the form of a Verilog or VHDL RTL description.

This is where the biggest problem lies. Finding a suitable architecture is not a simple task, and finding an optimal one is even more challenging. But the fundamental issue is the manual nature of this entire approach. As clever as we can be and no matter what we do, our curse, as engineers, is to trip over these tiny yet enormously frustrating things we've dubbed "bugs." Simply put, any manual intervention is a source of errors. Suddenly, what was initially a straightforward process from specification to implementation becomes a nightmarish iterative cycle. The hand-coded RTL design is tested, bugs are reported, and time is spent trying to hunt them down and fix them individually—only to move on to the next bug. This could be an endless process if it didn't have to end at some point to meet deadlines.

Keeping up with the pace

The issue of course is exacerbated by growing design sizes. The bigger the system and the more complex the application, the more chances of errors and the harder it becomes to stay on schedule. Unfortunately, ever-increasing complexity is one of the few certainties in electronic design.

Just remember the kind of equipment we had fifteen years ago, whether cell phones or televisions. Now compare them with today's commensurable items. Their evolution has been so dramatic that we don't even call them the same thing anymore! We now have "smart phones" in our pockets and "high-definition home entertainment systems" in our living rooms. Everything has changed to become more sophisticated, more complex. Likewise in the electrical engineering: technology nodes and process geometries keep shrinking, clock frequencies keep increasing, embedded cores keep multiplying, and verification methodologies are borrowing object-oriented concepts from the software community. Everything has changed.

Everything but one thing: the RTL creation process. We are trying to develop 4G broadband modems with tools and methods inherited from the mid-90s, when GSM was the hot topic. We are trying to create H264 decoders with languages adopted to design VGA controllers. Something is deeply broken. We simply can't create RTL efficiently enough; eventually, bugs trigger and problems fire during the verification phase. It is no surprise if verification is now the bottleneck in any ASIC project.

Benefits of high-level synthesis

High-level synthesis addresses the root cause of this problem by providing an error-free path from abstract specifications to RTL. By using HLS, design teams greatly accelerate design time while also reducing the overall verification effort.

Reducing design and verification efforts

When working at a high-level of abstraction a lot less detail is needed for the description. For instance, at the functional level, engineers do not need to worry about implementation details such as hierarchy, processes, clocks, or technology. They are free to focus only on the desired behavior. This makes the description much easier to write. With fewer lines of code, the risk of errors is greatly reduced, and with fewer things to test for in the source, it is easier to exhaustively verify the model.

After the high-level model is written and verified, HLS automates the RTL implementation process. But if HLS tools eliminate manual interventions and errors, they do not eliminate engineering intervention. That is, decisions still need to be made. With high-level synthesis, engineers remain in control; they make the decisions and the HLS tool implements them. They simply have a more efficient and productive way of getting their job done. For instance, the designer decides upon the proper level of parallelism for an optimal architecture and constrains the HLS tool accordingly. In turn, the tool takes care of allocating and scheduling the needed hardware resources, building the datapath and control structures to produce a fully functional and optimized implementation. With HLS, correct RTL is obtained more rapidly, shortening the creation phase. In turn, the debug overhead is lowered and the verification burden is reduced.

More effective reuse

Working at a higher level of abstraction has an additional benefit. The design sources are now truly generic and therefore more versatile. For years, IP and reuse have been promoted as ways to address the design complexity challenge. But these strategies find their limits. RTL views describe what happens between two clocks edges. By definition this is tied to a specific technology and clock frequency. If retargeting legacy RTL is often possible, it is usually done at the expense of power, performance and area. Moreover making small changes to an existing IP to create a derivative can quickly turn into a much bigger project than anticipated. In contrast, when working with purely functional specifications, there are no such details as clocks, technology or micro-architecture in the source. This is information added automatically during the high-level synthesis process. And if new functionality is added to the IP, changes can be made and verified more easily in the abstract source and without the fear of breaking a pipeline or having to rewrite a state machine. With HLS it is much simpler to reuse and retarget functional IP.

Investing R&D resources where it really matters

There are many other advantages to using high-level synthesis, but what is especially interesting is to look at the induced benefits. When properly used, HLS flows can help save months of R&D effort. With engineering resources spending fewer cycles on RTL coding and verification, more time can be spent on differentiating activities. RTL coding is a necessity, not a value-added activity. In comparison, algorithm development, architecture optimization, and system-level power optimization can really make a difference in the success of a product. Time-to-market often matters, but it is just one part of the equation. Feature superiority, cost competitiveness, and power consumption are also critical success factors. By using HLS, organizations can spend less effort dealing with mundane design tasks and invest more intelligence where it matters most.

Seizing the opportunity

High-level synthesis is not a new idea. The promise of designing in a better way is as old as EDA itself. The evolution towards higher abstractions is rooted in EDA's DNA. The industry constantly strives to raise the abstraction level, easing the design process for engineers around the world. When moving from transistor to gates, and then from gates to RTL, we did nothing other than adopt more efficient and higher-level hardware design methods. Today, once more, the design pressure is too high to resist the call for change.

Since the early commercial and academic work, HLS has come of age. A new generation of C synthesis tools reached the market in 2004. Since then, countless user testimonials and hundreds of tape-outs have confirmed not only the viability but also the necessity of HLS for modern ASIC design. Over the past few years, HLS tools have developed and added the necessary technology to become truly production-worthy. Initially limited to datapath designs, HLS tools have now matured to address complete systems, including control-logic, and complex SoC interconnects—without a penalty in quality of results.

The value of HLS has clearly been established and the technology routinely delivers on the expectations. High-level synthesis provides great benefits, but is also a disruptive technology. It implies change in the methodologies, in the design processes, and to some extent, in the skills required. The learning curve is the last barrier to wider adoption. The move to HDL languages didn't happen overnight either. Designers learned from books, references materials, and real-world examples, earning their RTL know-how over many years. The same is happening now for high-level synthesis. Early adopters have anchored HLS in their design flows and are paving the way for mainstream users.

This book will help designers travel this HLS road. It is meant to be a practical and valuable companion for engineers seeking to adopt high-level synthesis. The HLS promise awaits, the technology delivers it, and this book helps you seize and implement this necessary and more productive path to verified RTL.

Thomas Bollaert, January 2010

Chapter 2

General C++ Style

Introduction

The purpose of this C++ synthesis style guide is to provide a firm foundation for writing good quality synthesizable C++ code. This includes not only recommendations for achieving good quality of results in hardware, but also good programming practices to ensure "clean" code that passes compilation, execution, and RTL/C++ co-verification.

File Organization

This style guide covers general coding guidelines, including how to organize and structure the files that make up a design. This is only intended as an example recommendation and users are free to choose and use any structure that is comfortable, or required by their institution. The main intent here is to guide the user to adopting and adhering to a methodology that makes managing their designs easier.

An example directory structure for organizing your C++ files along with Catapult project files should look something like:

```
|---Project directory
            |----src
            |----ccs
            |----dat
            |----sim
```

Where:

- "Project directory" is the current design directory
- "src" directory contains all C++ (*.cpp, *.cxx, *.C, *.h, *.hpp) source and header files and the Makefile. This is where the executable is compiled and linked
- "ccs" contains the synthesis *.tcl scripts

- "dat" directory contains any file I/O for the testbench
- "sim" directory is for Matlab and Simulink projects and scripts

Building an Executable Using Makefiles

Make is a Unix utility that is used to automate the compilation of a set of files into an executable. Although it is not necessary to use Makefiles, it is highly recommended, and streamlines the compilation and linking process. Make has default rules it knows about and enforces, such as understanding file dependencies and guaranteeing that files are recompiled when a dependency changes. (For a more complete guide to creating makefiles see http://www.gnu.org/software/make/manual)

Makefile Naming

The "make" utility looks for a file called "Makefile" by default. If this file is not found it then looks for a file called "makefile". You can also specify an arbitrary filename by using the "-f" command line switch for Make.

Comments

Comments are denoted by preceding text with the pound (#) sign. Any text following the "#" sign untill the end of the line is treated as a comment.

```
# Example Makefile
```

Macros

Macros can be defined in a Makefile that allow substitution of complex expressions. For example:

```
CXX = /usr/bin/g++
```

Targets

The basic makefile is composed of a set of rules.

```
targets : prerequisites
command
```

The targets are file names and must be separated by spaces. The command lines must start with a tab character, and the prerequisites, also known as dependencies, consist of file names separated by spaces. The dependencies are used to test when a target is out of date.

Phony Targets

A phony target is simply a way to enforce commands to be executed regardless of whether a file of the same name exists in the Makefile directory. Thus the target will always execute even if a file of the same name is up to date.

```
.PHONY: clean
clean: rm *.o *.exe
```

Simple Makefile Example

Consider the simple example where the design consists of three files:

- hello.cpp - src directory file with a function that prints hello
- hello.h - include directory file that contains the function prototype
- main.cpp - testbench that calls the hello.cpp function.

A very basic Makefile to compile these files into an executable is shown in Example 2-1.

Example 2-1. Simple Makefile

```
# Example Makefile

#MACROS
CXX = /usr/bin/g++
CXXFLAGS = -O

#my_tb target is dependent on main.o and hello.o
my_tb : main.o hello.o
    ${CXX} ${CXXFLAGS} -o  my_tb main.o hello.o

#main.o is dependent on main.cpp and hello.h
main.o : main.cpp hello.h
    ${CXX} ${CXXFLAGS} -c main.cpp

#hello.o is dependent on hello.cpp and hello.h
hello.o : hello.cpp hello.h
    ${CXX} ${CXXFLAGS} -c hello.cpp
```

```
#phony target to remove all objects and executables
.PHONY: clean
clean:
    rm -f *.o *.exe
```

An improvement of the very simple Makefile shown in Example 2-1would be to take advantage of the use of macros and also the implicit dependency in Make between *.o files and *.cpp or *.cxx files. This is shown in Example 2-2.

Example 2-2. Makefile Using Macros

```
# Example Makefile

#MACROS
CAT_HOME    = $(MGC_HOME)
TARGET      = my_tb
OBJECTS     = main.o hello.o
DEPENDS     = hello.h
INCLUDES    = -I"$(CAT_HOME)/shared/include"
DEFINES     =
CXX         =  /usr/bin/g++
CXXFLAGS    = -g  -o3 $(DEFINES) $(INCLUDES)

$(TARGET): $(OBJECTS)
    $(CXX) $(CXXFLAGS) -o $(TARGET) $(OBJECTS)

$(OBJECTS): $(DEPENDS)

#phony target to remove all objects and executables
.PHONY: clean
clean:
    rm -f *.o *.exe
```

Header/Include Files

In C++ programming a header file typically contains forward declarations of classes, function prototypes, and other information shared by multiple source files. Although variables can be declared within a header file, making them global, this is not recommended. One of the most common high-level synthesis uses for header files is for creating type and constant definitions. Example 2-3 shows the header file for the hello.cpp example discussed in the previous section.

Example 2-3. Header File

```
//guard string to prevent multiple inclusion
#ifndef __HELLO__
#define __HELLO__
```

```
4
5  //Forward declaration of function
6  void hello();
7
8  const int ITERATIONS = 22;
9
10 typedef int dType;
11
12 #endif
```

The details of Example 2-3 are:

- Lines 2 and 3 implement a guard string that prevents multiple inclusion of the header file. The first time the header file is compiled "__HELLO__" will be defined, preventing further inclusion.
- Line 6 defines the function prototype for the "hello" function implemented in hello.cpp. Including this header file within another design makes the "hello" function available.
- Line 8 defines a constant integer "ITERATIONS" and sets it equal to 22. This could also have been done using a #define but it is not recommened.Use of #define should only be used when absolutely necessary. Defining constants using #define can lead to cryptic errors during compilation if the user is not careful.
- Line 10 uses a type definition to define a new type "dType" to be type int. This is very useful in that it allows the design data types to be decoupled from the implementation code. This mechanism can be used to easily switch between data types.

Example 2-4 shows the hello.cpp design which includes the header file.

Example 2-4. Including the Header File

```
#include <iostream>
using namespace std;
//Including user defined header file
#include "hello.h"
void hello(){
  //dType defined in header file
  dType tmp;

  //ITERATIONS defined in header file
  for(int i=0;i<ITERATIONS;i++){
    tmp = i;
    cout << "Hello" <<  tmp << endl;
  }
}
```

The details of Example 2-4 are:

- Line 4 includes the design header file
- Line 10 uses the constant "ITERATION" defined in the header file.

Example 2-5 shows the design testbench for the hello.cpp design. This file also includes the header file which gives it access to the function prototype which is instantiated on line 4.

Example 2-5. Testbench

```
1   //Include user header file
2   #include "hello.h"
3   int main(){
4       hello();
5   }
```

Test Benches

The user testbench is a C++ design that is used to test the device under test (DUT) for functional correctness. In a HLS design environment the C++ testbench is typically used to test both the C++ and the RTL, so it is important to follow good programming practices. Furthermore it is critical to leverage the C++ testbench to prove that the DUT matches the original algorithm as code changes are made. There is nothing worse than re-writing your C++ code to get good synthesis results only to find out that you have broken the functionality. Be smart, be methodical, and verify your design at every step.

Note

If you don't have a C++ testbench, write one. Otherwise you're wasting valuable time.

Creating a Golden Reference Design

One of the first things a new HLS user discovers is that they have to make code changes to their original floating or fixed point source code. These code changes are made to improve quality of results (QofR) and/ or pass synthesis. The biggest mistake that users can make is to take their algorithmic C++ code and start modifying it for synthesis without having created a backup reference to compare the changes against. It only takes a few code changes to completely break a design.

Consider the following design shown in Example 2-6:

Example 2-6. C++ with Conditional Branches

```
#include "test.h"
void test(dType a[2], dType b[2], dType dout[2], bool sel){

  if(sel){
    dout[0] = a[0] + b[0];
    dout[1] = a[1] + b[1];
  }else{
    dout[0] = a[0] - b[0];
    dout[1] = a[1] - b[1];
  }
}
```

Examination and/or synthesis of the "test" function shown in Example 2-6 reveals that the IO accesses on a, b, and dout creates a performance bottleneck if the interfaces are memory interfaces (The reasons for this performance bottleneck is discussed in later chapters). Rewriting the code allows for better performance, but rather than modifying the original code, a new design is created which allows a comparison to the original algorithm.

Note

Always make a copy of the original algorithm to verify against any code changes when possible. It may not be possible to do a bit-for-bit comparison for some algorithms.

The re-written design, which is functionally equivalent to the original, is shown in Example 2-7.

Example 2-7. Modified Design

```
#include "test_mod.h"
void test_mod(dType a[2], dType b[2], dType dout[2], bool sel){

  for(int i=0;i<2;i++){
    if(sel){
      dout[i] = a[i] + b[i];
    }else{
      dout[i] = a[i] - b[i];
    }
  }
}
```

The testbench should be modified to check the modified design against the original algorithm, shown in Example 2-8.

Example 2-8. Modified Testbench

```
1  #include <iostream>
2  using namespace std;
3  #include “test.h”
4  #include “test_mod.h”
5  int main(){
6    dType a[2] = {10, 20};
7    dType b[2] = {10, 20};
8    dType dout[2];
9    dType dout_mod[2];
10   bool sel = true;
11   bool error = false;
12   //DUT original
13   test(a,b,dout,sel);
14   //DUT modified
15   test_mod(a,b,dout_mod,sel);
16
17   for(int i=0;i<2;i++){
18     if(dout_mod[i] != dout[i]){
19       cout << “ERROR” << endl;
20       error = true;
21     }
22     else
23       cout << dout[i] << endl;
24   }
25   if(error)
26     return -1; //indicates test failure
27   else
28     return 0; //test passed
29 }
```

The details of the modified testbench shown in Example 2-8 are:

- Lines 13 and 15 instantiate the original and modified functions and apply the same set of inputs. Each function produces its own outputs “dout” and “dout_mod”.
- Lines 17 through 24 check each of the outputs from the original design against the modified design to see if they match. If there is a mismatch it’s flagged as an error.
- Lines 25 through 28 check to see if any errors occured and return the test status.

Note

Each time a code change is made the testbench should be rerun to check the change against the original design. Failure to do this may mean hours of debugging to figure out which change broke the design.

Make Sure You're Fully Testing the DUT

One of the most common, and costly, mistakes users make when testing the DUT is failing to test all possible conditional branches based on the control inputs into the DUT. This can often lead to discovering functional differences between the DUT and the golden reference after having made significant changes required for synthesis. This is illustrated by looking at the testbench and DUT shown in Example 2-8. The testbench only ever tests the DUT with "sel = 1". "sel" is responsible for selecting one of two possible conditional branches in the DUT. One of the primary reasons for making code changes is when the design cannot be synthesized, or when timing, performance, or area must be improved. If any of these reasons require the code to be rewritten, it is likely that it will force the user to modify all branches of any conditions in the design. The user would be unaware if a functional mistake was introduced in the branch for "sel = 0" after rewriting the code. The testbench should be rewritten as:

Example 2-9. Improved Testbench

```
#include <iostream>
using namespace std;
#include "test.h"
#include "test_mod.h"
int main(){
  dType a[2] = {10, 20};
  dType b[2] = {10, 20};
  dType dout[2];
  dType dout_mod[2];
  bool sel = true;

  for(int j=0;j<2;j++){
    sel = j;
    //DUT original
    test(a,b,dout,sel);

    //DUT modified
    test_mod(a,b,dout_mod,sel);

    for(int i=0;i<2;i++){
      if(dout_mod[i] != dout[i])
    cout << "ERROR" << endl;
      else
    cout << dout[i] << endl;
    }
  }
}
```

Now the DUT is tested for both values of "sel" covering both conditional branches.

Uninitialized Variables

In general a variable should never be read before it is written. Uninitialized variables are treated differently by different compilers and synthesis tools, often leading to unpredictable results. Many hours can be wasted trying to track down simulation bugs only to discover that the source of the problem is an uninitialized variable. Another common side effect is to have entire sections of a design optimized away because variables are not initialized. Consider the following design:

Example 2-10. Uninitialized Variables

```
1   void acc(int din[4], int &dout){
2     int tmp;
3     for(int i=0;i<4;i++)
4       tmp += din[i];
5     dout = tmp;
6   }
```

Line 2 of Example 2-10 defines a variable “tmp” that is left uninitialized. Line 4 then uses “tmp” to accumulate the array “din”. Since “tmp” is not initialized it can be considered as a “don’t care”, which means that the first accumulate looks like “tmp = (don’t care) + din[i]”. This can lead to a unexpected result. Most compilers will flag this as a warning if verbose messaging is enabled. E.g. “g++ -v....””.

Note

Leaving variables uninitialized can cause unexpected results in both synthesis and simulation. Some designs may pass C++ simulation yet fail RTL simulation, leading to costly, yet necessary, debugging by the designer.

Chapter 3
Bit Accurate Data Types

Introduction

Algorithm designers, system architects, and RTL engineers have been using bit-accurate data types for years to model true hardware behavior. The need for bit-accuracy becomes especially obvious now that designers are building hardware directly from C++, whose native types only come in widths of 1, 8, 16, 32, etc, bits. Many existing bit accurate data types used today are "home grown" class libraries that evolved within companies, and model bit accuracy using traditional shift and mask techniques. Although these "home grown" types may be faster for simulation, they are typically very slow for synthesis, and can also give much poorer quality of results than the industry standard bit-accurate data types.

To date there are two industry standard bit accurate data types; the SystemCTM and Mentor Graphics Algorithmic C data types. Although SystemC was developed first, the implementation of its bit-accurate data types suffers from a number of issues, the biggest being long execution runtimes. Because of this, customer demand drove Mentor to develop their own bit-accurate types, which have now become the most widely used data types in high-level synthesis. The Algorithmic C data types not only simulate much faster than the SystemC types, but give better quality of results for synthesis over "home grown" bit accurate types. Algorithmic C data types are also consistent between C++ and RTL simulation. So whatever you build in C++ matches the true hardware behavior. In light of this, the focus of this chapter is on the use of the Algorithmic C data types. Furthermore, this chapter only attempts to provide enough of an overview of the Algorithmic C data types to begin designing in C++. A comprehensive manual is available at:

http://www.mentor.com/products/esl/high_level_synthesis/ac_datatypes

Compilation, Debug, and Simulation Speed

In order to compile and use the Algorithmic C data types the header file for either the integer data types, ac_int, or fixed point data types, ac_fixed, must be included in the C++ source file(s).

```
#include <ac_int.h>
#include <ac_fixed.h>
```

It is also critical to achieving the fastest runtimes that the highest level of optimization is set (-O3 in gcc and /Ox in MS Visual).

```
g++ -O3 -I <path to Algorithmic C data types> -o hello.exe hello.
cpp
```

When debugging bit accurate code using gdb, ddd, or any of the MS Visual tools it is best to disable optimizations and turn on verbose warnings. E.g.

```
g++ -g -Wall -I <path to Algorithmic C data types> -o hello.exe
hello.cpp
```

Header Files and Typedefs

Although the Algorithmic C data types execute much faster than the SystemC data types, they will in general run slower than the native C++ types. For this reason, as well as simplifying debugging, it is often desirable to be able to quickly switch between Algorithmic C and native C++ data types. The easiest way to do this is to define all variables in a global header file for both Algorithmic C and native types and use compiler defines to switch between the two definitions. Example 3-1 shows a header file that uses a compiler define "NATIVE_TYPES" to select between the type definitions of "dType" and "oType" as either native C++ types or Algorithmic C data types. This header file is then included in Example 3-2, which defines all its variables in terms of the type defined variables.

Example 3-1. Header File with Typedefs

```
#ifndef __TYPEDEFS__
#define __TYPEDEFS__
#include <ac_int.h>
```

```
#ifdef NATIVE_TYPES
typedef short int dType;
typedef int       oType;
#else
typedef ac_int<7,true> dType;
typedef ac_int<14,true> oType;
#endif
#endif
```

Example 3-2. Using Type Defined Variables in a Design

```
#include "typedefs.h"
void test(dType a, dType b, oType &c){
  oType tmp;

  tmp = a*b;
  c = tmp;
}
```

Integer Data Types

The Algorithmic C integer data types allow designers to model a signed or unsigned bit vector with static bit precision. This closely matches what RTL designers can do today with VHDL and Verilog 2001. The ac_int data types are templatized, and allow designers to specify both the width and signedness of variables.

Unsigned integer

The Algorithmic C unsigned integer data types are declared as:

```
ac_int<W,false> x;
```

where:

```
W = Bit width
0 <= x <= 2W-1 by increments of 1
```

Any value assigned to "x" that is either greater than the maximum representable value, or negative in this case, will overflow or "wrap" around. This is the same behavior that RTL designers are familiar with today when creating counters. Example 3-3, which is purely for simulation, shows the use of a 7-bit unsigned integer to create a sign wave.

Example 3-3. Algorithmic C Unsigned Integer

```
1   #include <ac_int.h>
2   #include <fstream>
3   #include <cmath>
4   using namespace std;
5   const double pi = 3.14;
6   const int OFFSET = 64;
7   int main(){
8     fstream fptr;
9     fptr.open(“tmp.txt”, fstream::out);
10    ac_int<7,false> x[128];
11    for(int i=0;i<128;i++){
12      x[i] = OFFSET + 63*sin(2*pi*i/64);
13      fptr << x[i] <<endl;
14    }
15    fptr.close();
16  }
17
```

The details of Example 3-3 are:

- Line 1 includes the ac_int library.
- Line 10 declares an array of 7-bit unsigned integers.
- Line 12 computes two cycles of a sine wave and assigns the results to “x[i]”. The sine wave amplitude is +/- 63 and it is given a positive offset of 64 to utilize the full dynamic range of “x”. This is because a 7-bit unsigned integer can range from 0 to 2^7-1 or 0 to 127. Since the sine function only produces values between 1 and -1 it is necessary to scale it and add the offset before assigning to “x”

The plot of the sine wave generated in Example 3-3 is shown in Figure 3-1.

Figure 3-1. Maximum Range of 7-bit Unsigned Integer

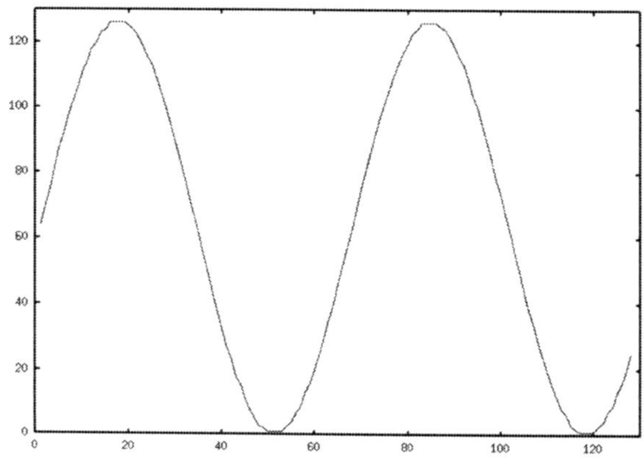

The effects of wrapping in a bit-accurate data type can be seen by replotting the results of Example 3-3 when the offset is increased to 80, shown in Figure 3-3.

Figure 3-2. Effect of Wrapping in Bit-accurate Unsigned Data Types

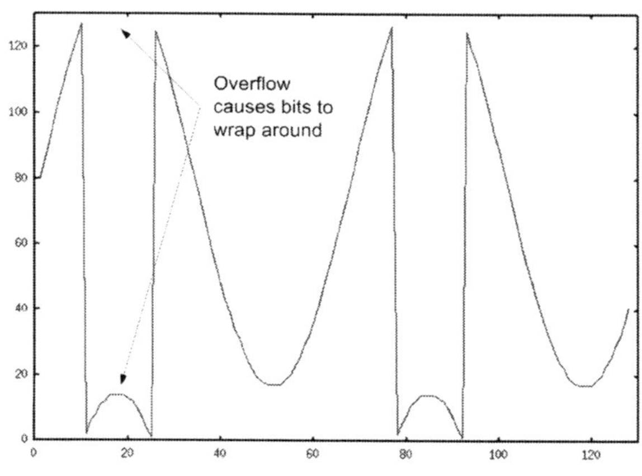

Signed Integer

The Algorithmic C signed integer data types are declared as:

```
ac_int<W,true> x;
```

where:

```
-2W-1<= x <= 2W-1-1 by increments of 1
```

The signed integer bit-accurate data types have similar wrapping behavior as the unsigned integers with the difference being that the signed data types wrap based on the expression shown above. Example 3-4 shows the sine wave generation example where the type for "x" has been changed to signed integer on Line 10. The offset has been set to zero since the negative values are now supported by the data type.

Example 3-4. Algorithmic C Signed Integer

```
1   #include <ac_int.h>
2   #include <iostream.h>
3   #include <fstream.h>
4   #include <math.h>
5   const double pi = 3.14;
6   const int OFFSET = 0;
7   int main(){
8     fstream fptr;
9     fptr.open("tmp.txt", fstream::out);
10    ac_int<7,true> x[128];
11
12    for(int i=0;i<128;i++){
13      x[i] = OFFSET + 63*sin(2*pi*i/64);
14      fptr << x[i] <<endl;
15    }
16    fptr.close();
17  }
```

Figure 3-3 and Figure 3-4 show the plots for Example 3-4 with offsets of zero and 14 respectively.

Figure 3-3. Maximum Range of a 7-bit Signed Integer

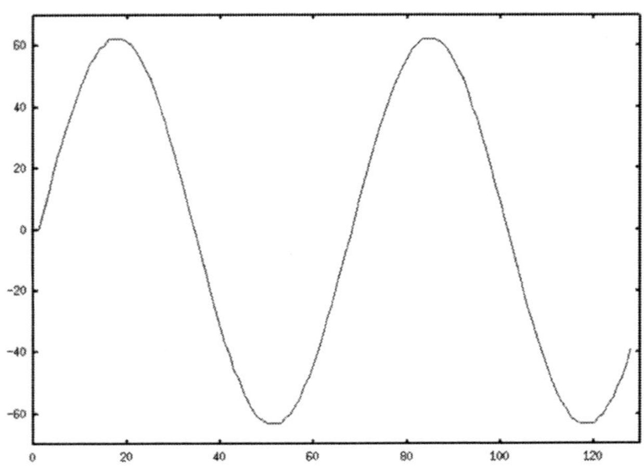

Figure 3-4. Effects of Wrapping in Bit-accurate Signed Data Types

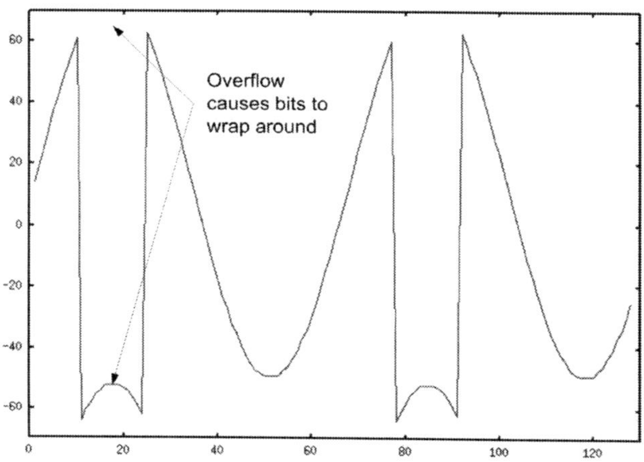

Note

The plots of both the unsigned and signed data types illustrate the similarity in behavior to designing in RTL. Designers MUST be aware of the dynamic range of the algorithm to avoid "wrapping" of bits. Once this occurs the algorithm results are meaningless.

Fixed Point Data Types

The Algorithmic C fixed point data types allow designers to model a signed or unsigned bit vector with static fixed point precision. This is something that cannot be done directly in RTL, and is one of the many advantages of high-level synthesis. Although most DSP algorithms are designed using floating or fixed point arithmetic, the actual RTL implementation is done using integers, and the designer has to manually track the decimal point by shifting intermediate results left or right. This is not only a tedious way of designing, but it is also error prone. HLS allows designers to build hardware directly from a fixed point model. The ac_fixed data types are templatized, and allow designers to specify both the integer and fractional width and signedness of variables.

Unsigned Fixed Point

The Algorithmic C unsigned fixed point data types are declared as:

```
ac_fixed<W,I,false> x;
```

where:

```
0 <= x <= (1-2-W)2I by increments of 2I-W
```

For the fixed point data types both the total width “W” and the number of integer bits “I” are specified. As a result of this “I” determines the location of the decimal point relative to the MSB (Figure 3-5).

Figure 3-5. Fixed Point Decimal Point Position

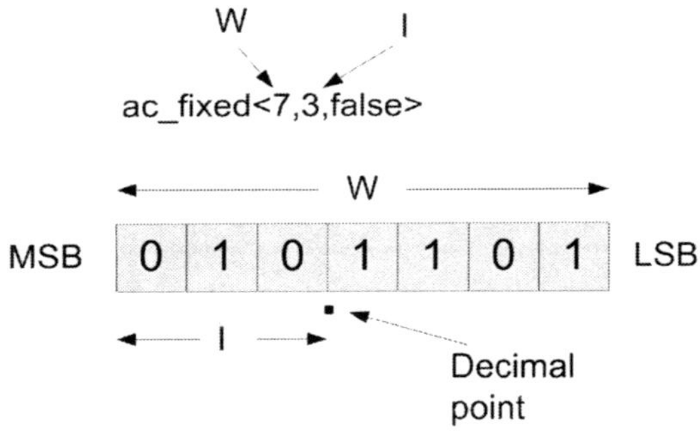

Now that the Algorithmic C data types have provided the ability to express fractional values, it is no longer necessary to scale results into the integer domain. Taking Example 3-3, which scaled the sine wave up to maximize the dynamic range of 7-bit unsigned integer “x’, and expressing it using unsigned fixed point data types leads to Example 3-5.

Example 3-5. Algorithmic C Unsigned Fixed Point Data Type

```
#include <ac_fixed.h>
#include <iostream.h>
#include <fstream.h>
#include <math.h>
const double pi = 3.14;
const double OFFSET = 1.0;
int main(){
  fstream fptr;
  fptr.open("tmp.txt", fstream::out);
  ac_fixed<7,1,false> x[128];
```

```
11
12     for(int i=0;i<128;i++){
13       x[i] = OFFSET + 0.98*sin(2*pi*i/64);
14       fptr << x[i] <<endl;
15     }
16     fptr.close();
17   }
```

The details of Example 3-5 are:

- Line 1 includes the ac_fixed data types.
- Line 10 defines a 7-bit fixed point array "x" with one integer bit. This means that values of "x" can range somewhere between 0 and (1-2-7)*21 or 1.98. Supporting the full amplitude of the sine wave from -1 to 1 would require 2 integer bits.
- Line 13 computes the sine wave and adds an offset of one, keeping all values positive and preventing wrapping by keeping the amplitude of the sign wave from exceeding +/- 0.98.

Figure 3-6 shows the plot of the sine wave from Example 3-5.

Figure 3-6. 7-bit Unsigned Fixed Point Sine Wave

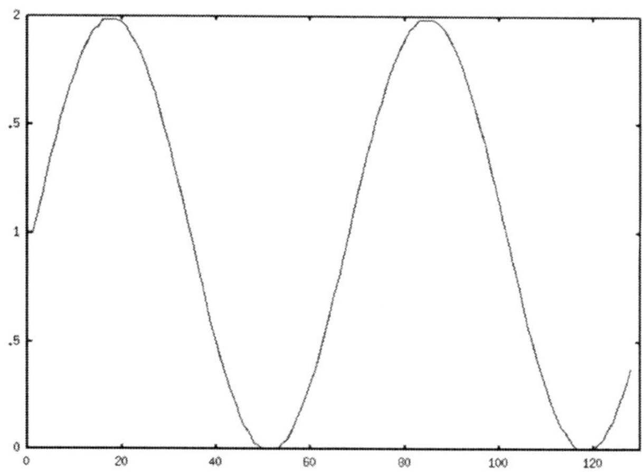

Signed Fixed Point

The Algorithmic C signed fixed point data types are declared as:

```
ac_fixed<W,I,true> x;
```

where:

```
-0.5*2I <= x <= (0.5-2-W)2I by increments of 2I-W
```

"I" determines the location of the decimal point relative to the MSB, which is also the sign bit.

Example 3-6 shows Example 3-5 rewritten to use signed fixed point data types. The offset, which is no longer needed is set to zero. Figure 3-7 shows the plot of the sine wave from Example 3-6, which ranges from almost -1 to 1.

Example 3-6. Algorithmic C Signed Fixed Point Data Type

```
#include <ac_fixed.h>
#include <iostream.h>
#include <fstream.h>
#include <math.h>
const double pi = 3.14;
const double OFFSET = 0.0;
int main(){
  fstream fptr;
  fptr.open("tmp.txt", fstream::out);
  ac_fixed<7,1,true> x[128];

  for(int i=0;i<128;i++){
    x[i] = OFFSET + 0.98*sin(2*pi*i/64);
    fptr << x[i] <<endl;
  }
  fptr.close();
}
```

Figure 3-7. 7-bit Signed Fixed Point Sine Wave

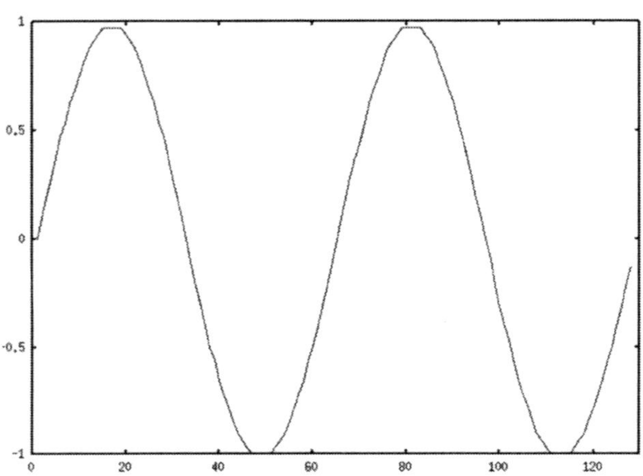

Quantization and Overflow

In addition to allowing algorithms to be expressed more naturally, the fixed point data types also provide mechanisms to deal with quantization and overflow. The default mode for ac_fixed is to truncate and wrap/overflow, similar to what was shown for ac_int. The default mode does not cost any additional area but may not be ideal for some applications. There are many quantization and overflow modes supported by the ac_fixed data types, and they are covered in detail in the Algorithmic C data types manual. This chapter presents the reasons why one might wish to enable these modes. The quantization and overflow modes are enabled using additional template parameters for the ac_fixed data types.

```
ac_fixed<W,I,S,Q,O> x;
```

Where Q and O set the quantization and overflow modes respectively.

Truncation and Rounding

The default behavior is truncate and to throw bits to the right of the LSB away. This results in a complete loss of information. An example of this would be assigning fractional data to a fixed point variable with only integer bits. E.g.

```
ac_fixed<7,7,true> x = 0.5;
```

Printing out the value of “x” after the assignment gives a value of zero, since “x” has no fractional bits. Instead of throwing away the fractional data, a rounding mode can be used to round up or down depending on the fractional value. Similar to what we were all taught in grade school, we can round up or down depending on where the fractional value lands between two integer values.

```
ac_fixed<7,7,true,AC_RND> x = 0.5;
```

AC_RND rounds up towards positive infinity, which means that “x” will be assigned a value of one. The rounding mode rounds based on the smallest allowable increment defined by W and I.

Saturation and Overflow

The previous examples showed that the default behavior is to have bits "wrap" around when the maximum or minimum representable value is exceeded. This is known as overflow or underflow and is usually a very undesirable situation. Most algorithms can, and should, be designed so that overflow never occurs. This means taking into account the dynamic range of variables and ensuring that the internal bit growth is sufficient to represent all possible ranges of algorithm inputs. However there are situations where it is necessary to ensure that overflow can never occur. Mission critical systems such as flight control would be a good example where overflow would be disastrous. Video algorithms are another example of why overflow would be undesirable, with most people not wanting pixels flipping from the brightest to the darkest colors.

Taking Example 3-6 on page 23 and adding a slight positive offset will cause the result to overflow, leading to a meaningless, or potentially catastrophic, result, shown in Figure 3-8.

Figure 3-8. Fixed Point Overflow

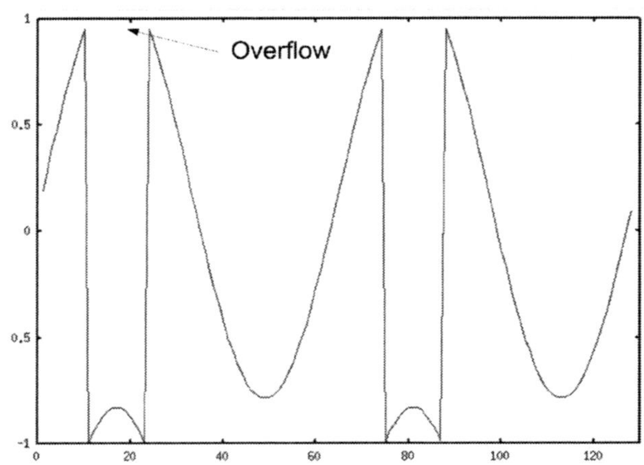

Overflow can be prevented by enabling saturation on the fixed point data type. More details on the behavior of these modes can be found in the Algorithmic C data types manual. Care should always be taken when using saturation since it often increases the area of a design. DO NOT simply turn on saturation on all variables in a design. Saturation is typically used selectively in a few places. Example 3-6 is rewritten to enable saturation using the overflow template parameter AC_SAT, shown on line 10 of Example 3-7.

Example 3-7. Turning on Saturation in ac_fixed Data Types

```
1   #include <ac_fixed.h>
2   #include <iostream.h>
3   #include <fstream.h>
4   #include <math.h>
5   const double pi = 3.14;
6   const double OFFSET = 0.2;
7   int main(){
8     fstream fptr;
9     fptr.open("tmp.txt", fstream::out);
10    ac_fixed<7,1,true,AC_TRN,AC_SAT> x[128];
11
12    for(int i=0;i<128;i++){
13      x[i] = OFFSET + 0.98*sin(2*pi*i/(double)64);
14      fptr << x[i] <<endl;
15    }
16    fptr.close();
17  }
```

Figure 3-9 shows the results of adding saturation in Example 3-7.

Figure 3-9. Effects of Saturation

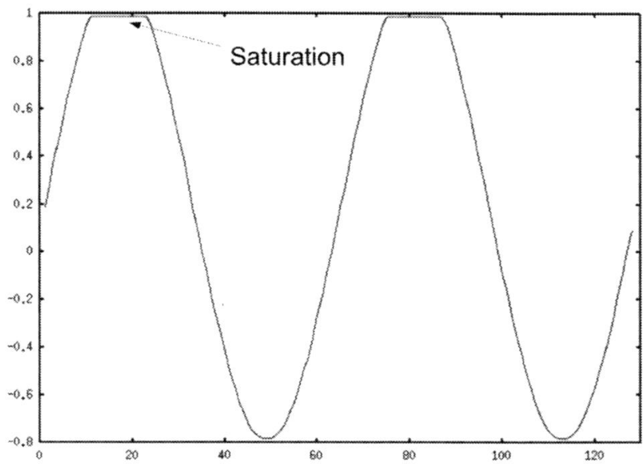

Examination of Figure 3-9 shows that enabling saturation has prevented any overflow or under flow. However the figure also clearly shows that the waveform is non-linear. This non-linearity, also known as clipping, has the effect of adding noise into the system. This is one of the main reasons why most algorithms should be designed without the use of saturation.

Note

Saturation should only be used when absolutely necessary. Most algorithms can be designed to avoid overflow/underflow by selecting the appropriate variable bit widths to support the full dynamic range of the algorithm. Saturation usually impacts both area and performance.

Operators

All of the standard C++ arithmetic and logical operators are supported by the ac_int and ac_fixed data types. The operators such as multiplication, addition, etc, are designed to return a result without a loss of precision. The Algorithmic C reference manual should be consulted for a full description of all operators. Operators such as divide "/" and modulus "%" are supported but should be used with care when the two operands are variables. This is because an operation such as division costs a great deal in area, and the reality is that most hardware designers would never use a hardware divider. This is often a misunderstood area of HLS since it is perfectly reasonable to write something like z = x/y in C++. If a divide is truly needed, there are cheaper implementations such as the CORDIC algorithm. Most HLS tools provide a library that implements these functions more efficiently. Divisions or modulus by a constant are much more acceptable, and are implemented using add and shift logic. Divide or modulus by a power of two is implemented using static shifts, and cost nothing in additional area. Two operators whose behavior is worth noting are the bit select and shift operators, but first a quick discussion of arithmetic operators.

Bitwise Arithmetic Operators: *, +, -, /, &, |, ^,%

The Algorithmic C bitwise arithmetic operators are designed so that there is no loss of precision in the return value. Furthermore the mixture of signed and unsigned Algorithmic C data types is supported, and returns the expected signedness.

The return type of an arithmetic operation automatically takes care of bit growth so that there is no loss of precision. This can be done automatically because the bit widths of the two operands are specified as template parameters when the ac_int variables are declared. This is shown below:

```
ac_int<8,true> a,b; //8-bits signed
```

Multiplying "a" times "b", each with 8 bits of precision, requires 16 bits of precision:

```
(returns ac_int<16,true>)(a*c)
```

Adding "a" plus "b" requires nine bits of precision:

```
(returns ac_int<9,true>)(a+b)
```

Bit Select Operator: []

Individual bits can be read or written from an ac_int or ac_fixed data type using the [] operator. The operator index selects the bit position. E.g. x[1], x[3], x[7]. The return value is an object of class ac_int::bitref and a built-in conversion function to ac_int and bool are provided. The code fragment below shows how the bit select operator can be used to read the sign bit of an ac_int.

```
ac_int<11,true> x;
bool is_neg;

if(x[10]) //test for sign bit treated as bool
    is_neg = true;
else
    is_neg = false;
```

The bit select operator can just as easily be used to write a bit in an ac_int or ac_fixed. E.g.

```
ac_fixed<9,1,false> x = 0;
bool add_one = true;

if(add_one)
    x[8] = 1; //set MSB of x
```

Shift Operators: <<, >>

The shift operators are worthy of discussion because, unlike the arithmetic operators which maintain full precision, they return the precision of the left operand. This can lead to unexpected results. Furthermore, designers should pay attention to how they use shifts because both operands can be signed or unsigned. This can also lead to surprising results that would not be possible in traditional RTL.

Shift Right Operator: >>

Unsigned Shift Right

The unsigned right shift operator applied to unsigned ac_int or ac_fixed data types behaves exactly how an RTL designer would expect a shifter to behave. Each bit is shifted right by the shift amount and zeros are stuffed into the upper bits. This can best be seen by a graphical example shown in Figure 3-10. The variable being shifted is a 4-bit unsigned integer that is initialized with all bits set equal to 1. As the shift amount is increased zeros are stuffed into the MSBs.

```
ac_int<4,false> x = -1; //set all bits to 1's
int idx;
ac_int<4,false> y = x >> idx;
```

Figure 3-10. Unsigned Shift Right

y	idx	Value of y
1 1 1 1	0	15
0 1 1 1	1	7
0 0 1 1	2	3
0 0 0 1	3	1
0 0 0 0	4	0

Signed Shift Right

Signed shift right has somewhat unexpected behavior. Most hardware engineers think of right and left shifts as either divides or multiplies by power of two. So one would expect that at some point right shifting sets all bits to zero as was shown in Figure 3-10. However, when right shifting a signed ac_int, the sign bit is always kept, which has the end result of shifting ones from the MSB rather than zeros. This is shown in Figure 3-11. The ac_int "x" is initialized with the sign bit set equal to one and all other bits zero, which is equal to negative eight. Each increasing shift has the effect of dividing by increasing powers of two until all bits are set to one. From this point forward the result is always negative one.

```
ac_int<4,true> x = 0;
x[3] = 1 //set x equal -8
int idx;
ac_int<4,true> y = x >> idx;
```

Figure 3-11. Signed Right Shift

y	idx	Value of y
1 0 0 0	0	-8
1 1 0 0	1	-4
1 1 1 0	2	-2
1 1 1 1	3	-1
1 1 1 1	4	-1

Shift Left Operator: <<

The left shift operator behaves mostly as one would expect, with the exception being when assigning to a variable with larger precision. The typical behavior is discussed first.

Unsigned Shift Left

Shifting an unsigned ac_int left and assigning the result to a variable with the same precision has similar behavior as an unsigned shift right, except that zeros are stuffed from the LSB position, shown in Figure 3-12.

```
ac_int<4,false> x = -1; //set all bits to 1's
int idx;
ac_int<4,false> y = x << idx;
```

Figure 3-12. Unsigned Shift Left

y	idx	Value of y
1 1 1 1	0	15
1 1 1 0	1	14
1 1 0 0	2	12
1 0 0 0	3	8
0 0 0 0	4	0

Signed Shift Left

Signed shift left has similar behavior to unsigned shift left where zeros are stuffed from the LSB.

```
ac_int<4,true> x = -1; //set all bits to 1's
int idx;
ac_int<4,true> y = x << idx;
```

Figure 3-13. Signed Shift Left

y	idx	Value of y
1 1 1 1	0	-1
1 1 1 0	1	-2
1 1 0 0	2	-4
1 0 0 0	3	-8
0 0 0 0	4	0

Unexpected Loss of Precision

Shifting left can have unexpected, but correct, behavior when the expectation is that the result is based on the precision of the target variables. This is best understood by looking at an example:

```
ac_int<4,false> x = -1; //set all bits to 1's
int idx;
ac_int<8,false> y = x << idx;
```

In the example shown above, "x", which is four bits unsigned, is shifted left and assigned to "y", which is eight bits unsigned. A common misconception by designers new to Algorithmic C data types is that the upper bits of "x" are stored in "y" as they are shifted past the MSB of "x". Figure 3-14 shows the actual behavior. Remember that the shift operator returns the precision of the left operand, which is four bits in this example. In order to preserve the bits that are being shifted out of the MSB of "x" it is necessary to cast "x" to the precision of "y". E.g.

```
ac_int<4,false> x = -1; //set all bits to 1's
int idx;
ac_int<8,false> y = (ac_int<8,false>) x << idx;
```

Note

Shifting left can result in loss of bits if the variable being shifted is not cast to the same precision as the left hand of the assignment.

This now gives the desired behavior shown in Figure 3-15.

Figure 3-14. Unexpected Loss of Precision Shifting Left

y								idx	Value of y
0	0	0	0	1	1	1	1	0	15
0	0	0	0	1	1	1	0	1	14
0	0	0	0	1	1	0	0	2	12
0	0	0	0	1	0	0	0	3	8
0	0	0	0	0	0	0	0	4	0

Figure 3-15. Casting to Desired Precision Shifting Left

y								idx	Value of y
0	0	0	0	1	1	1	1	0	15
0	0	0	1	1	1	1	0	1	30
0	0	1	1	1	1	0	0	2	60
0	1	1	1	1	0	0	0	3	120
1	1	1	1	0	0	0	0	4	240

Methods

The Algorithmic C data types provide a number of built-in methods which are covered in detail in the reference manual. The methods most often used for design and simulation are covered below.

Slice Read: slc

The slice read method has the form: slc<W>(int lsb)

Where the template parameter "W" specifies the width of the slice, and "lsb" points to where the slice begins. Dynamic sizing of the slice width is not possible because "W" is a template parameter. However it is possible to dynamically change the value of "lsb". Consider the following example where a three bit slice is read from "x" and assigned to "y". "lsb" points to bit position two. Figure 3-16 shows graphically how the read slice works.

```
ac_int<8,false> x = 100;
ac_int<3,false> y;

y = x.slc<3>(2);
```

Figure 3-16. Read Slice

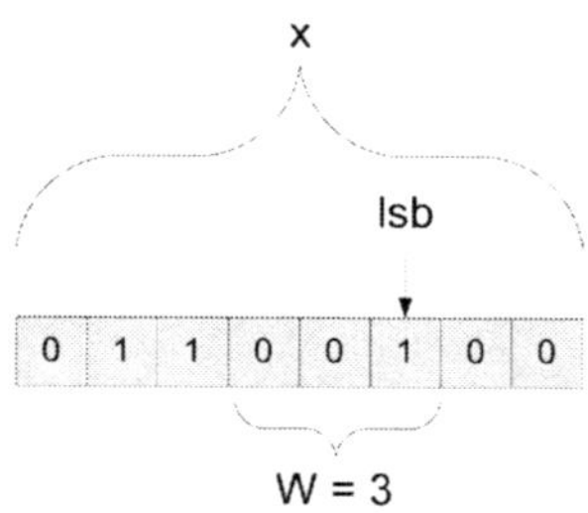

Problems with Compilation of Read Slice Method

Some compliers may error out when the read slice method is used inside of a templatized function or class. If this occurs simply place the keyword "template" before the method name. E.g.:

```
ac_int<8,false> x = 100;
```

```
ac_int<3,false> y;

y = x.template slc<3>(2);
```

Slice Write: set_slc

The set_slc method has the form: set_slc(int lsb, const ac_int<W,S> &slc)

Where "lsb" is the index into the bit vector and indicates where the slice should be written. This is essentially the same behavior as that shown for slice reads in Figure3-16. The slice that is written can be a signed or unsigned ac_int. Similar to slice reads, the size of the slice cannot be changed dynamically. The usage is:

```
ac_int<8,false> x = 0;
ac_int<3,false> y = 5;

x.set_slc(2,y);
```

Explicit Conversion Functions

The Algorithmic C data types provide a number of implicit, as well as explicit, conversion functions. These are fully documented in the language reference manual. The explicit conversion functions are typically required when assigning an ac_fixed to an ac_int, or when trying to use "printf" to print out simulation results. The first case generates a compiler error, and the second case using "printf" causes a segmentation fault during runtime.

E.g.

```
ac_fixed<8,3,false> x = 3.185;
ac_int<8,false> y;
```

The assignment below generates a compiler error

```
y = x;
```

It should be written as:

```
y = x.to_int();
```

The statement below causes a runtime segmentation fault:

```
printf("%d\n",y);
```

It should be rewritten as:

```
printf("%d\n",y.to_int());
```

Implicit Conversion Functions

The algorithmic C data types have built-in conversion functions for the assignment operator "=" up to 64-bits. These conversion functions automatically convert algorithmic C data types back to native types. This allows variables to be used directly for conditional testing of any bits set in the bit vector. E.g.

```
ac_int<64,false> ctrl = 37;
if(ctrl)//test for any bit set to a one
    tmp += din;
```

Note

A compilation error occurs if "ctrl" is greater than 64 bits. In this case one of the explicit conversion functions should be used.

Helper/Utility Functions

The Algorithmic C libraries provide some useful functions for designing hardware.

Array Uninitialization: ac::init_array

This function can be used to both initialize and uninitialize an array with a constant or a don't care value. The advantage of using this built-in function is that the HLS tool can optimize the array more efficiently since it doesn't have to analyze loops and assignments to determine the intent. The most common use of this function is to uninitialize an array, or in other words to set all array elements to don't-care. The reason why one would wish to do this is in order to prevent the creation of hardware to clear all locations of a static array mapped to memory. By definition in the C++ language a variable declared as static is set equal to zero when it is constructed. In many cases it is not necessary to clear the memory of a design since it is known that the memory is written before it is ever read. Thus the overhead of initially cycling through each memory location can be removed. The ac::init_array function has the form:

```
bool ac::init_array<"init value">("base address of array", "number
of elements");
```

Consider the following code fragment where the array "a" is mapped to a memory.

```
static int a[1000];
static bool dummy = ac::init_array<AC_VAL_DC>(a,1000);
```

Note

The reason why "dummy" is used and declared static is that we only want to call the init_array function once during initialization.

ceil, floor, and nbits

It is often required to statically compute the minimum number of bits required for an Algorithmic C data type in order to index an array. The following functions are provided:

- ac::log2_ceil<x>::val - returns the number of bits required to index "x" elements.
- ac::log2_floor<x>::val - returns log2(x).
- ac::nbits<x>::val - returns number of bits required to represent "x".

In all of these cases "x" must be statically determinable. For example:

```
const int WORDS = 175;
void foo(int data[WORDS],
    ac_int<ac::log2_ceil<WORDS>::val,false> idx,
    int &dout){

    dout = data[idx];
}
```

In the example above "idx" has been defined with the minimum number of bits required to index "WORDS" elements.

Complex Data Types

The Algorithmic C bit accurate data types libraries also provide support for complex data types, eliminating the need for designers to develop their own complex class. The ac_complex data type is a templatized class that can be used with both ac_int and ac_fixed as well as native C++ types. The Algorithmic C data types manual should be referred to for usage and restrictions.

Chapter 4
Fundamentals of High Level Synthesis

Introduction

One of the common misconceptions held by people is that synthesizing hardware from C++ provides users the freedom of expressing their algorithms using any style of C++ coding that they desire. When designing using high-level C++ synthesis, it is important to remember that we are still describing hardware using C++, and a "poor" description can lead to a sub-optimal RTL implementation. It is the responsibility of the user to code their C++ to not only account for the underlying memory architecture of a design, but to also adhere to the recommended coding style for C++ synthesis. Because of this it is important to have a solid understanding about what high level synthesis really does. This chapter attempts to cover the basics of high level synthesis, and to show what designers can expect from a given coding style. Where appropriate, the code examples are accompanied by hardware diagrams to hopefully allow RTL designers to relate C++ synthesis to concepts that are familiar to them.

The Top-level Design Module

Similar to RTL design, HLS requires that users specify where the "top" of their design is. This is where the design interfaces with the outside world and consists of port definitions, direction, and bit widths or in the case of C++, data types. Since we are still designing hardware, although using C++, it is helpful to see how this might relate to the world of RTL design. Consider the following simple Verilog RTL design that describes a d-type flip flow with asynchronous reset.

Example 4-1. Simple Verilog Design Module

```
module top(clk,arst,din,dout);

input clk;
input arst;
input [31:0] din;
output [31:0] dout;
```

```
reg [31:0] dout;

always@(posedge clk or posedge arst)
begin
  if(arst == 1'b1)
    dout = 1'b0;
  else
    dout = din;
end

endmodule
```

The verilog module shown in Example 4-1 contains several input ports and a single output port. The inputs are for clock, reset, and 32-bit input data, the output is for the 32-bit output data. The port directions and widths are explicitly defined in the Verilog source.

In order for HLS to determine the top-level design from the C++, the user must either specify a pragma in the source code or set a user constraint in the synthesis tool.

So the C++ equivalent description of the Verilog design described above would look like:

Example 4-2. Setting the Top-level Design

```
void top(int din, int& dout){
  dout = din;
}
```

Even this simple example illustrates how HLS can simplify the design process. The C++ description is very compact. Looking at the C++ description in Example 4-2 it is important to understand that there are several things that are implied in the code.

Registered Outputs

High-level synthesis by default builds synchronous designs. This means that all outputs of the top-level design are registered to guarantee that timing is met when connecting to another design. There are mechanisms to build smaller combinational blocks (automated component flows) but in general designs are synchronous.

Control Ports

The C++ code has no concept of timing so there are no clocks, enables, or resets described in the source. Instead these signals are added by the synthesis tool. Control over things like polarity, type of reset, etc, are taken care of by setting design constraints.

Port Width

For the simple case, meaning minimal interface constraints in synthesis, the bit widths of the top-level ports, excluding clock and reset, are implied by the data type. In the design example shown above the data type is "int" which implies 32 bits. Designers can describe any arbitrary bit width using bit-accurate data types.

Port Direction

The port direction is implied by how an interface variable is used in the C++ code

Input ports

An input port is inferred when an interface variable is only read. In the C++ example shown in Example 4-2 you can see that "din" is only ever read, so it is determined to be an input. If a variable is declared on the interface as "pass by value" it can only be an input. This is covered in more detail later.

Output ports

An output port is inferred in two cases. One is when the top-level function returns a value. The other is when the interface variable is only written in the C++ code. In Example 4-2 it can be seen that "dout" is only ever written. It can also be seen that "dout" is declared as a reference. A variable must be declared as a reference or a pointer in order to infer an output. This is also covered in more detail later.

Inout Ports

Although this design does not contain any inout ports, these are inferred if an interface variable is both read and written in the same design. This requires that the variable is declared as a reference or a pointer.

High-level C++ Synthesis

Although this style guide is not intended to be a tutorial on the intricacies of high-level synthesis optimizations, it is useful to present a brief overview of the basic process of automatically transforming un-timed algorithmic C++ into hardware. Understanding these fundamental concepts goes a great way towards providing a solid foundation for the material covered in later sections.

Similar to the rest of the style guide, these concepts are best illustrated using simple examples consisting of both C++ code and hardware design concepts. Consider the following C++ example that accumulates four integer values shown in Example 4-3.

Example 4-3. Simple C++ Accumulate

```
#include "accum.h"
void accumulate(int a, int b, int c, int d, int &dout){
  int t1,t2;

  t1 = a + b;
  t2 = t1 + c;
  dout = t2 + d;
}
```

Data Flow Graph Analysis

The process of high-level synthesis starts by analyzing the data dependencies between the various steps in the algorithm shown above. This analysis leads to a Data Flow Graph (DFG) description shown in Figure 4-1.

Figure 4-1. Data Flow Graph Description

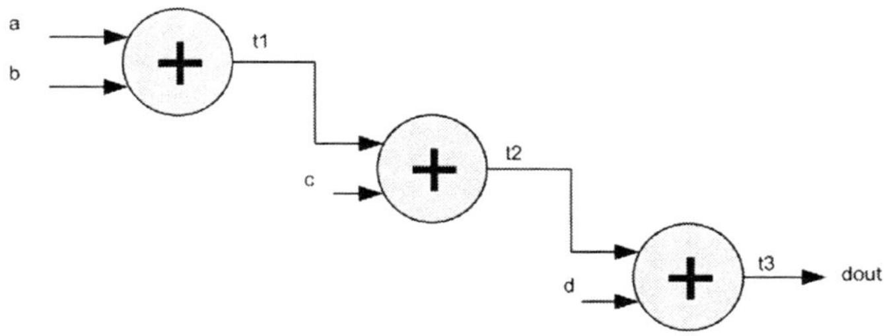

Each node of the DFG represents an operation defined in the C++ code, for this example all operations use the "add" operator. The connection between nodes represents data dependencies and indicates the order of operations. This example shows that t1 must be computed before t2 [1].

Resource Allocation

Once the DFG has been assembled, each operation is mapped onto a hardware resource which is then used during scheduling. This is the process known as resource allocation. The resource corresponds to a physical implementation of the operator hardware. This implementation is annotated with both timing and area information which is used during scheduling. Any given operator may have multiple hardware resource implementations that each have different area/delay/latency trade-offs. The resources are selected from a technology specific pre-characterized library that contains sufficient data points to represent a wide range of bit widths and clock frequencies. Figure 4-2 shows the resource allocation of Figure 4-1. Each operation can potentially be allocated to a different resource. It is also typical that designers can explicitly control resource allocation to insert pipeline registers or limit the number of available resources.

Figure 4-2. Resource Allocation

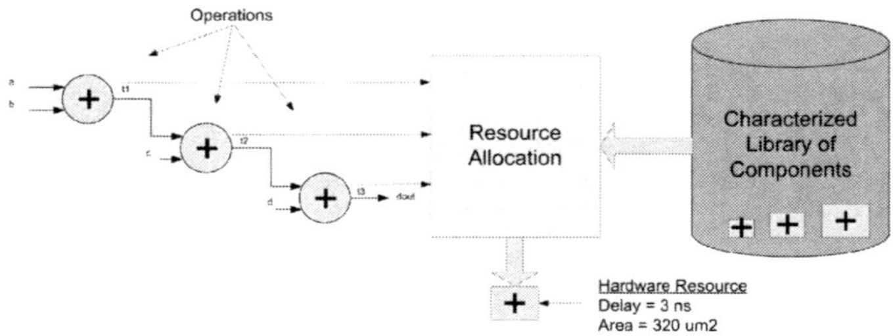

Scheduling

High-level synthesis adds "time" to the design during the process known as "scheduling". Scheduling takes the operations described in the DFG and decides when (in which clock cycle) they are performed. This has the effect of adding registers between operations based on a target clock frequency. This is similar to what RTL designers would call pipelining, by which they mean inserting registers to reduce combinational delays. This is not the same as "loop pipelining" which is discussed later.

If we assume that the "add" operation for the DFG of Figure 4-1 takes 3 ns out of a 5 ns clock cycle, the resulting schedule would look something like the schedule shown in Figure 4-3. Each add operation is scheduled in its own clock cycle C1, C2, and C3. Thus registers are inserted automatically between each adder.

Figure 4-3. Scheduled Design

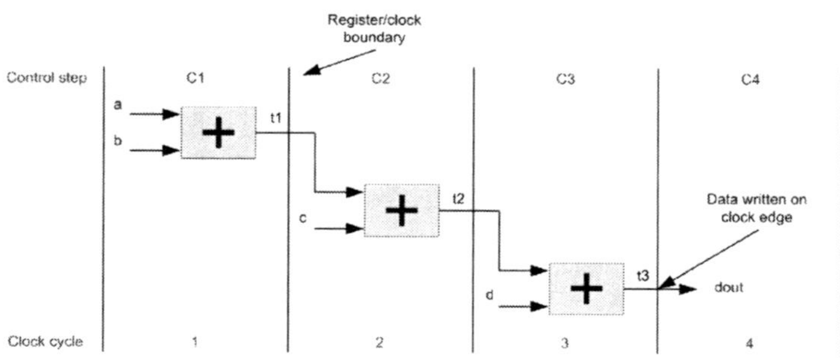

A data path state machine (DPFSM) is created to control the scheduled design. The FSM for this example requires four states that correspond to the four clock cycles needed to execute the schedule shown above. In HLS these state are also referred to as control steps or c-steps.

Figure 4-4. Data Path State Diagram

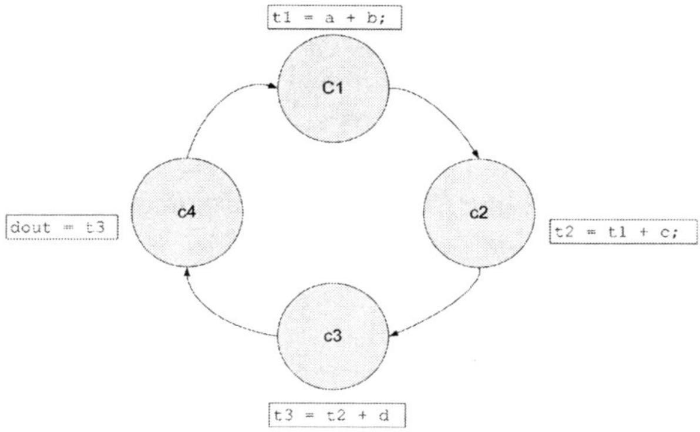

The state diagram of the DPFSM is shown in Figure 4-4 and illustrates that the scheduled design is capable of producing a new output every four clock cycles. Once C4 has completed a new C1 begins.

The resulting hardware that is generated from the schedule shown in Figure 4-3 varies depending on how the design was constrained in terms of resource allocation as well as the amount of loop pipelining used on the design. Loop pipelining is covered in detail next but for now let's assume that the design is unconstrained, which should allow the hardware to be realized with the minimum number of resources if sharing saves area. In this example that would be the minimum number of adders. The resulting hardware would look something like that shown in Figure 4-5.

Figure 4-5. Hardware Implementation of the Unconstrained Design

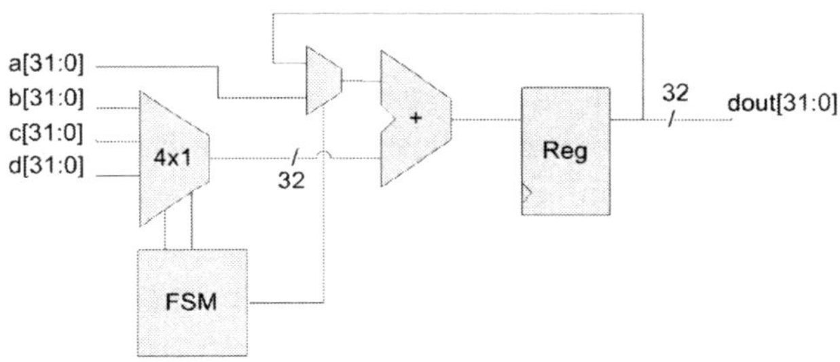

The resulting hardware shown in Figure 4-5 uses a single adder to accumulate a, b, c, and d. It should be noted that the data path is 32-bits wide because the variables have been declared as integer types.

Classic RISC Pipelining

The HLS concept of "Loop Pipelining" is similar to the classic RISC pipeline covered in most introductory computer architecture classes.

The basic five stage pipeline in a RISC architecture typically consists of Instruction Fetch(IF), Instruction Decode(ID), Execute(EX), Memory access(MA), and Register write back(WB) stages.

Figure 4-6. Five Stage RISC Pipeline

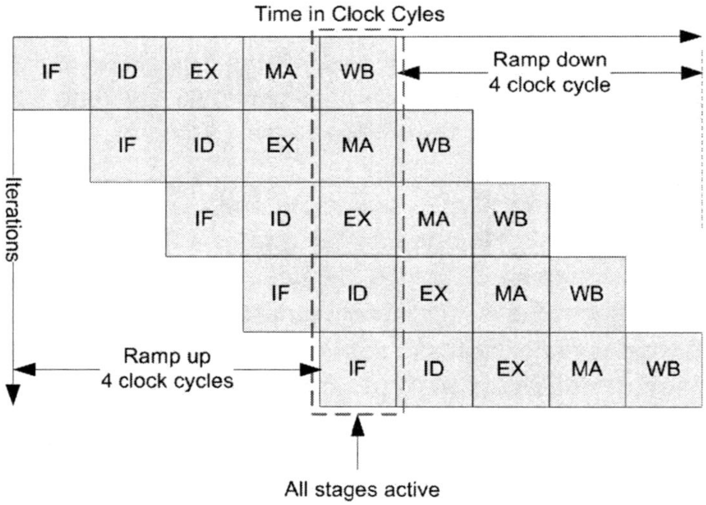

Figure 4-6 illustrates how a new instruction can be fetched each clock cycle while the other pipeline stages are gradually activated. The time it takes for all pipeline stages to become active is known as the pipeline "ramp up". Once all pipeline stages are active the pipeline "ramp down" is the time it takes for all pipeline stages to become inactive. The difference between the RISC pipeline and HLS loop pipelining is that the RISC pipeline is designed to fetch and execute every clock cycle. A design that does not need to run every clock cycle under-utilizes the pipeline, and a design that needs to fetch and execute multiple times per clock cycle is not possible. HLS removes these restrictions and allows the pipeline to be custom built to meet the design specification.

Loop Pipelining

Similar to the RISC pipelining example described in Figure 4-6, which allows new instructions to be read before the current instruction has finished, "Loop Pipelining" allows a new iteration of a loop to be started before the current iteration has finished. Although in Example 4-3 there are no explicit loops, the top-level function call has an implied loop, also known as the main loop. Each iteration of the implied loop corresponds to execution of the schedule shown in Figure 4-3 on page 39. "Loop pipelining" allows the execution of the loop iterations to be overlapped, increasing the design performance by running them in parallel. The amount of overlap is controlled by the "Initiation Interval (II)". This also determines the number of pipeline stages

Note

The Initiation Interval (II) is how many clock cycles are taken before starting the next loop iteration. Thus an II=1 means a new loop iteration is started every clock cycle.

The initiation interval is set on a desired loop either as a design constraint in the HLS design environment, or alternatively can be set using a C++ compiler pragma.

Note

Latency refers to the time, in clock cycles, from the first input to the first output

Note

Throughput, not to be confused with IO throughput, refers to how often, in clock cycles, a function call can complete.

If the design of Example 4-3 on page 38 is left unconstrained there is only a single pipeline stage because there's no overlap between execution of each iteration of the main loop. This results in data written every four clock cycles (Figure 4-7). The design has a latency of three and a throughput of four clock cycles. Because there is no overlap of any operation only a single adder is required if sharing reduces overall area.

Figure 4-7. No Pipelining, L=3, TP=4

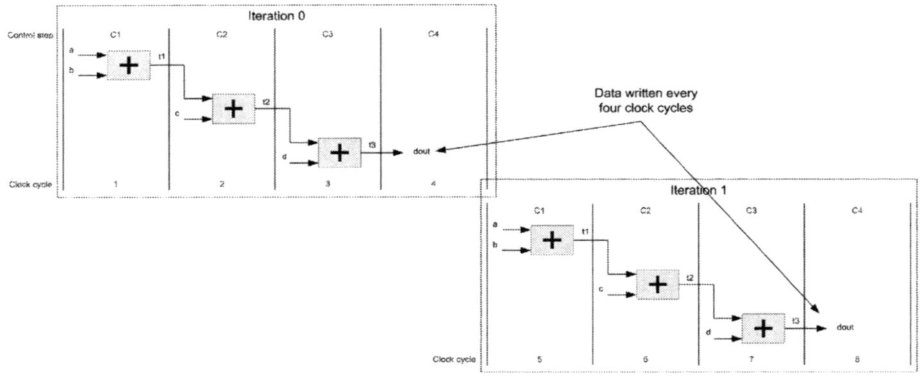

If a pipelining constraint of II=3 is applied on the top-level design (main loop), then the next loop iteration can be started in C4 allowing writing of "dout" in C4 to be overlapped with the reading of the next values of "a" and "b". The output is now written every three clock cycles while still requiring only one adder to implement the hardware (Figure 4-8). Only one pipeline stage is required since C4 is only used to allow the completion of the write on "dout".

Note

The number of pipeline stages increases by one if other operations are scheduled in the last c-step.

Figure 4-8. Pipeline II=3, L=3, TP=3

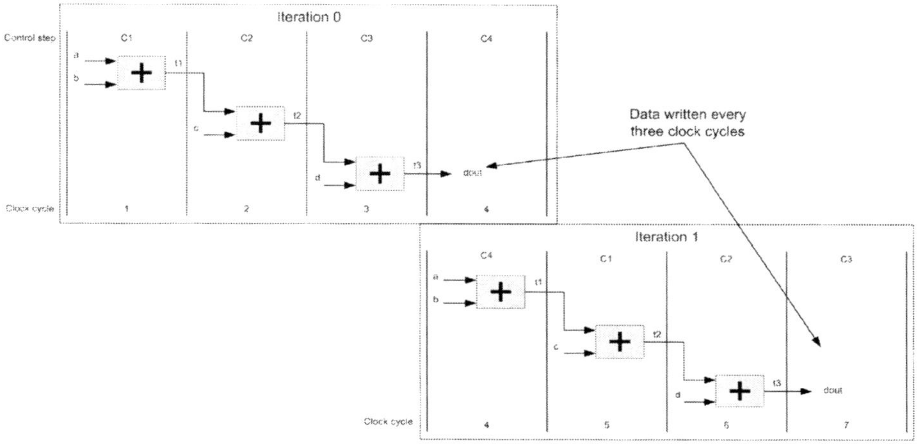

Figure 4-9 shows that pipelining with an II=2 results in a new iteration started every two clock cycles. Iteration one is started in C3 while iteration 0 is computing "t3 = t2 + d". Since iteration one is computing "t1 = a + b" it can be seen that two adders are required for the two pipeline stages.

Figure 4-9. Pipeline II=2, L=3, TP=2

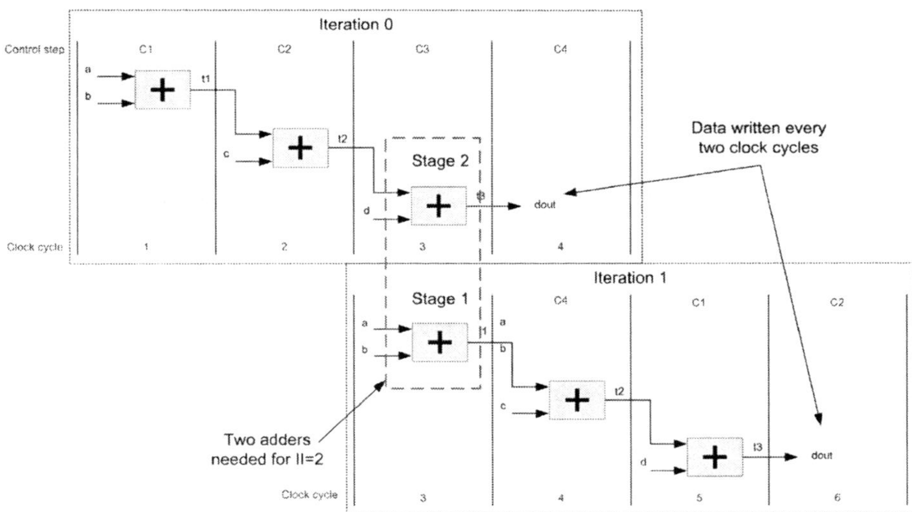

Pipelining with an II=1 (Figure 4-10) results in a new iteration started every clock cycle. Iteration one is started in C2 and iteration 2 is started in C3. Looking at C3 in the Figure 4-10 shows that three adders are required in hardware since all three pipeline stages are active.

Figure 4-10. Pipelining II = 1, L=3, TP=1

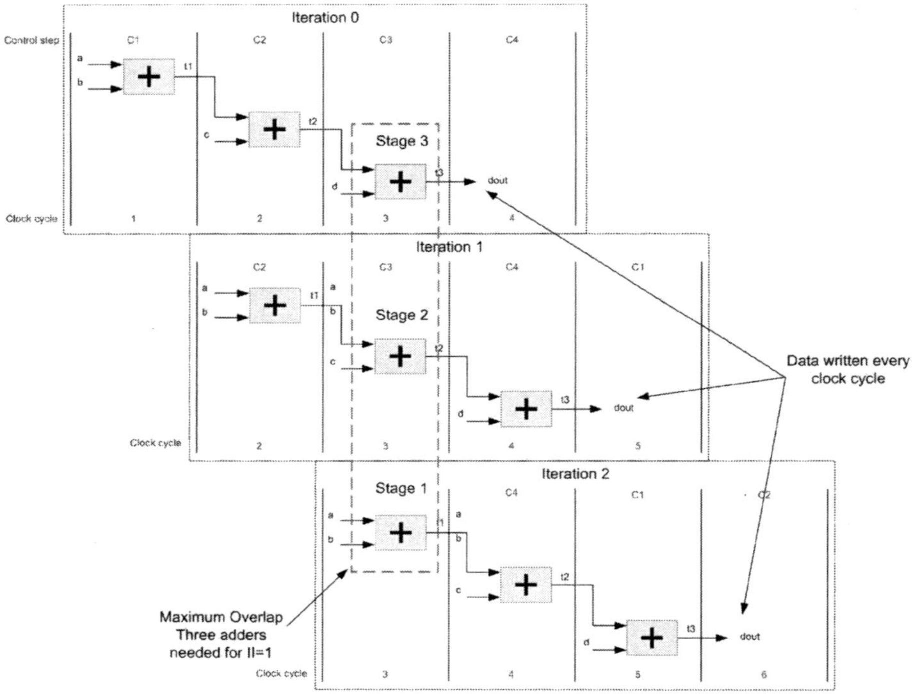

Loops

One of the most important features of HLS for tuning design performance is Loop Unrolling. However, it is necessary first to discuss what constitutes a "loop" in C++. Loops are the primary mechanism for applying high level synthesis constraints as well as moving data, or IO, into and out of an algorithm. The style in which loops are written can have a significant impact on the quality of results of the generated hardware. In order to talk about how to write loops it's helpful to introduce a few definitions:

- Interface synthesis - the process of mapping top-level C++ variables to resources that implement an interface protocol (wire, handshake, memory).
- Loop iterations - the number of times the loop runs before it exits.
- Loop iterator - the variable used to compute the loop iteration.
- Loop body - the code between the start and the end of the loop.
- Loop unrolling - the number of times to copy the loop body.
- Loop pipelining - how often to start the next iteration of the loop.

What's in a Loop?

In HLS a design always has one loop which corresponds to the top-level function call. This is known as the "main loop" (See line 1 of Example 4-4).

Example 4-4. The Main Loop

```
1  void top(int din, int& dout){
2    dout = din;
3  }
```

The "main loop" is a continuously running loop, which means that it runs for an infinite number of iterations. The analog to this can be seen from the equivalent Verilog module implementation that was shown in the top-level interface section. Once that Verilog module is reset it runs forever as long as the clock is supplied.

There are three ways to specify a loop in C++; using the "for" loop, "while" loop, and "do-while" loop. The syntax is as follows:

"for" Loop

Syntax:

```
LABEL: for( initialization; test-condition; increment ) {
    statement-list or loop body;
 }
```

The “for” construct is a general looping mechanism consisting of 4 parts:

1. initialization - which consists of zero or more comma-delimited variable initialization statements
2. test-condition - which is evaluated to determine if the execution of the for loop continues
3. increment - which consists of zero or more comma-delimited statements that increment variables
4. statement-list - which consists of zero or more statements that execute each time the loop is executed.

Example 4-5. “for” Loop

```
#include "simple_for_loop.h"
void simple_for_loop(int din[4], int dout[4]){
 FOR_LOOP:for(int i=0;i<4;i++){
    dout[i] = din[i];
  }
}
```

The example “for” loop shown in Example 4-5 copies four 32-bit values from din to dout. The for loop has initialization “int i=0”, test condition “i<4”, and increment “i++”.

“while” Loop

The “while” keyword is used as a looping construct that executes the loop body as long as condition is tested as true. If the condition starts off as false, the loop body is never executed. (You can use a do loop to guarantee that the loop body executes at least once.)

Syntax:

```
LABEL: while(test-condition) {
    statement-list or loop body;
 }
```

Example 4-6. “while” Loop

```
#include "simple_while_loop.h"
void simple_while_loop(int din[4], int dout[4]){
  int i=0;
 WHILE_LOOP:while(i<4){
    dout[i] = din[i];
```

```
        i++;
      }
    }
```

The “while” loop shown in Example 4-6 has the same functionality of the previous “for” loop example.

“do” Loop

The “do” keyword is used as a looping construct that executes the loop body until the condition is tested as false. The loop body always executes at least once.

Syntax:

```
LABEL: do{
    statement-list or loop body;
 } while(test-condition);
```

Example 4-7. “do” Loop

```
#include "simple_do_loop.h"
void simple_do_loop(int din[4], int dout[4]){
  int i=0;
 DO_LOOP:do{
    dout[i] = din[i];
    i++;
  }while(i<4);
}
```

The “do” loop shown in Example 4-7 has the same functionality as the previous “for” and “while” loop examples.

Rolled Loops

Note

If a loop is left “rolled”, each iteration of the loop takes at least one clock cycle to execute in hardware. This is because there is an implied “wait until clock” for the loop body.

Consider the following C++ example that uses a “for” loop to accumulate four 32-bit integers from an array:

Example 4-8. C++ Accumulate Using Loops

```
void accumulate4(int din[4], int &dout){
  int acc=0;
 ACCUM:for(int i=0;i<4;i++){
    acc += din[i];
  }
  dout = acc;
}

Design Constraints
Main loop pipelined with II=1
All loops left rolled
```

Although the loop is left rolled notice that the design has been pipelined with an II=1. This was done intentionally in order to ignore the effects of the extra clock cycle required for allowing the write of "dout" to complete, as it was discussed in "Loop Pipelining" on page 41. The effects of pipelining loops is covered in more detail in later sections. Figure 4-11 shows the schedule of the loop iterations for Example 4-8.

Figure 4-11. Schedule for Accumulate Using Loops

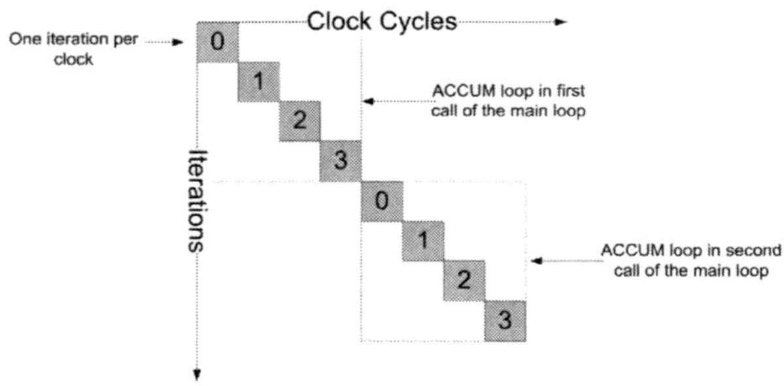

Figure 4-11 shows that each call to the "accumulate" function requires four clock cycles to accumulate the four 32-bit values in Example 4-8. This is because the loop has been left rolled and there is an implied "wait until clock" at the end of the loop body. The synthesized hardware would have the approximate structure shown in Figure 4-12

Figure 4-12. Hardware Implementation - Accumulate Using Loops

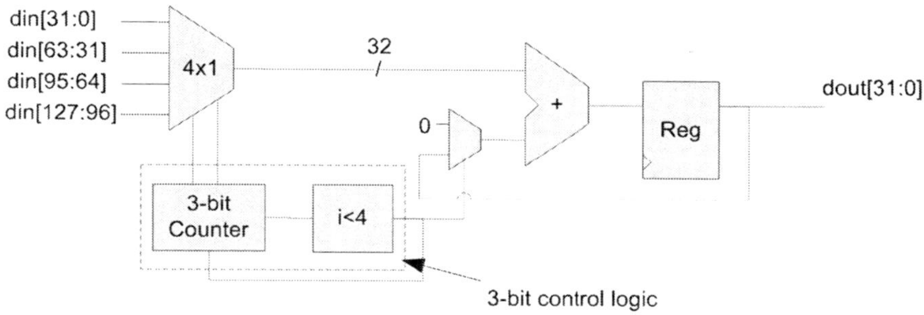

Figure 4-12 has a structure similar to what one might expect from a hand-code RTL design. However, one important feature to note is that the control logic for this implementation is three bits wide. The reason for this is that the loop exit condition is “i<4”. This means that this loop only exits when “i>=4”, which requires at least three bits.

Note

The number of bits required for evaluating the loop exit condition is usually one bit larger than expected. This is because the loop iteration must increment before exiting.

Loop Unrolling

Loop unrolling is the primary mechanism to add parallelism into a design. This is done by automatically scheduling multiple loop iterations in parallel, when possible. The amount of parallelism is controlled by how many loop iterations are run in parallel. This is different than loop pipelining, which allows loop iterations to be started every II clock cycles. Loop unrolling can theoretically execute all loop iterations within a single clock cycle as long as there are no dependencies between successive iterations.

Partial Loop Unrolling

If we take Example 4-8 and unroll the ACCUM loop by a factor of two, this has the equivalent effect of manually duplicating the loop body two times and running the ACCUM loop for half as many iterations. Example

4-9 illustrates the effects of loop unrolling by showing the ACCUM loop of Example 4-8 manually unrolled two times.

Example 4-9. Manual Loop Unrolling - Unroll by 2

```
1   void accumulate(int din[4], int &dout){
2     int acc=0;
3    ACCUM:for(int i=0;i<4;i+=2){
4       acc += din[i];
5       acc += din[i+1];
6     }
7     dout = acc;
8   }
```

The details of Example 4-9 are:

- line 3 increments the ACCUM loop by two, which means that the "partially unrolled" loop now has two iterations.
- Lines 4 and 5 have duplicated the loop body two times, which shows that two accumulations are performed each iteration. It should be noted that the accumulate in Line 5 is dependent on the accumulate on line 4. For now it is assumed that there is still sufficient time to schedule both in the same clock cycle. Dependencies between loop iterations are discussed later.

Figure 4-13 shows the schedule of the loop iterations for Example 4-8 on page 47 when the ACCUM loop is unrolled by two. All four values are now accumulated in only two clock cycles.

Figure 4-13. Schedule for Accumulate Unroll by 2

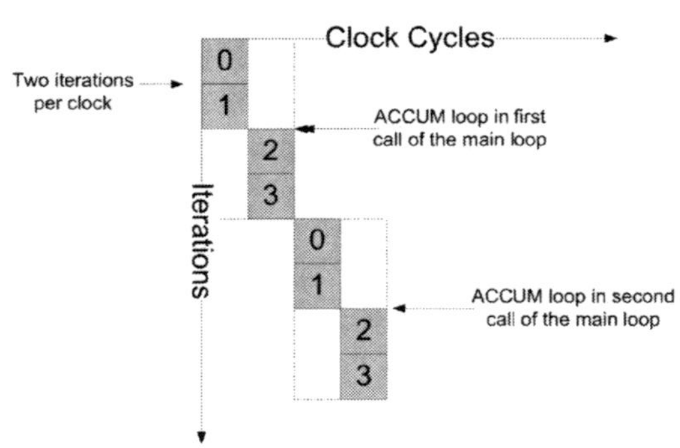

Figure 4-14 shows the hardware implementation when unrolling by two. It can be seen that this design requires twice as many resources (adders) as the "rolled" version.

Figure 4-14. Hardware Implementation - Accumulate Unroll by 2

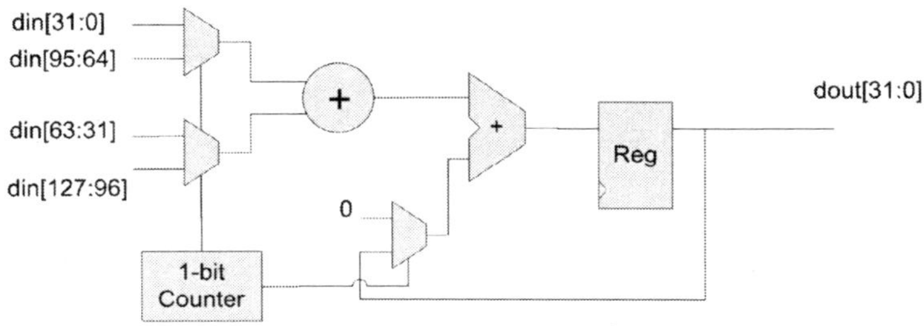

Fully Unrolled Loops

Taking Example 4-8 on page 47 and "fully" unrolling the ACCUM loop dissolves the loop and allows all iterations to be scheduled in the same clock cycle (Assuming that there is sufficient time to account for dependencies between iterations). The manual equivalent C++ of doing this is shown in Example 4-10.

Example 4-10. Manual Loop Unrolling - Fully Unrolled

```
void accumulate(int din[4], int &dout){
    int acc=0;

    acc += din[0];
    acc += din[1];
    acc += din[2];
    acc += din[3];

    dout = acc;
}
```

Figure 4-15 shows the schedule of the fully unrolled ACCUM loop. All four values are now accumulated in a single clock cycle.

Figure 4-15. Schedule for Accumulate - Fully Unrolled

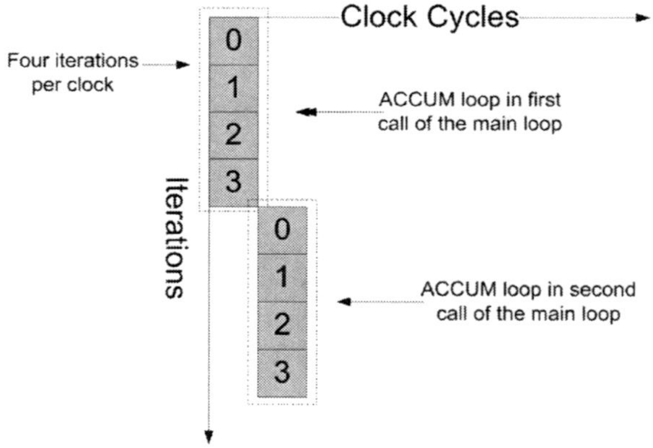

Figure 4-16 shows the approximate hardware when fully unrolling the ACCUM loop.

Figure 4-16. Hardware Implementation - Accumulate with Fully Unrolled Loop

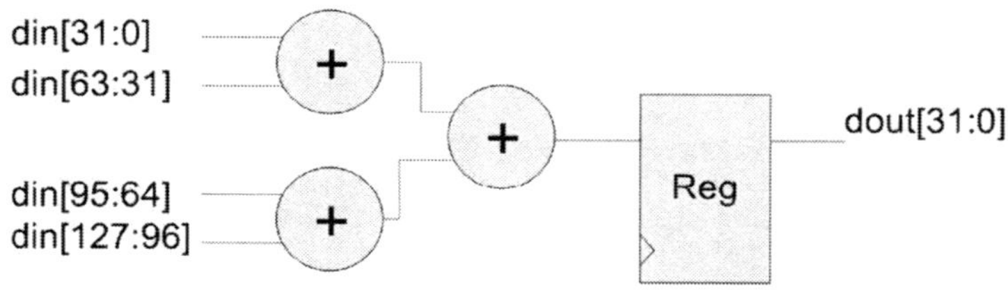

Dependencies Between Loop Iterations

Note

Unrolling a loop does not necessarily guarantee that the loop iterations are scheduled in the same c-step. Dependencies between iterations can limit parallelism.

The previous examples have assumed that there is sufficient time to ignore the effects of any dependencies between loop iterations. Thus Figure 4-15 shows all four iterations scheduled in the same clock cycle, but it does not show the dependencies that exist between iterations. A more accurate depiction of the schedule that includes the dependencies and component delays is shown in Figure 4-17. If the adders in Figure 4-17 were sufficiently slow it would be likely that second stage of the adder tree would be scheduled in the next clock cycle, increasing the design latency. However, if the design is pipelined with II=1 it is still possible to achieve a throughput of accumulating four values per clock cycle. Thus some dependencies between loop iterations do not limit design performance. However in many cases the dependencies between iterations limit performance or prevent pipelining. This is covered in detail in "Data Feedback" on page 73.

Figure 4-17. Schedule Dependencies

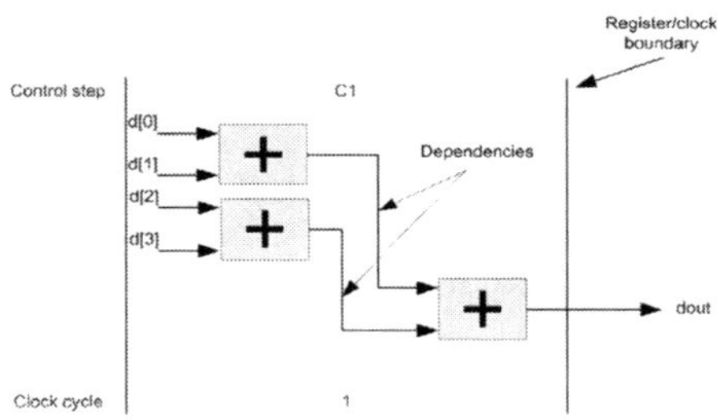

Loops with Constant Bounds

When writing loops for HLS it is important, when possible, to express them such that there is:

1. A constant initialization of the loop iterator
2. A test condition of the loop iterator against a constant value
3. A constant increment of the loop iterator

Writing the loop in this fashion allows HLS to optimize the design by reducing the bit widths of control and data path signals that are based on the loop iterator. This is because the three conditions listed above are sufficient for determining the maximum number of loop iterations. This is desirable to be able to get accurate information about latency and throughput of a design.

The four cycle accumulator in Example 4-8 on page 47 is a good example of writing loops with constant bounds. The corresponding hardware implementation shown in Figure 4-12 on page 48 shows that the control logic is optimally reduced to three bits. The main point to take away from this example is that even though the loop iterator "i" was declared as a 32-bit integer, HLS is able to reduce the bit widths to the fewest possible bits because the loop was written with constant bounds.

Loops with Conditional Bounds

The previous section showed that optimal hardware can be inferred if a loop is written with unconditional bounds. However, it is often the case that an algorithm or design requires that a loop terminate early based on some variable that has been defined outside of the loop, or on the design interface. This is a perfectly reasonable thing to do, but the way this is written in the C++ code can have a dramatic impact on the quality of results as well as accurate reporting of latency and throughput.

The accumulator design used in Example 4-8 can be modified to illustrate the impact in quality-of-results when using a loop with conditional bounds. In order to make the accumulator more programmable the code is modified so that the accumulator can accumulate anywhere from one to four 32-bit values.

Example 4-11. Conditional Accumulate

```
1   #include "accum.h"
2   #include <ac_int.h>
3   void accumulate(int din[4], int &dout, unsigned  int ctrl){
4     int acc=0;
5    ACCUM:for(int i=0;i<ctrl;i++){
6       acc += din[i];
7     }
8     dout = acc;
9   }
10
```

The modified accumulator design shown above now uses the interface variable "ctrl" on line 5 to select the number of loop iterations to be one through four. Synthesizing this design reveals that there are several inefficiencies with the resulting hardware.

Caution

Having a variable as the loop upper or lower bound often results in the loop counter hardware being larger than needed

Caution

Having a variable as the loop upper bound requires one extra clock cycle to test the loop condition

Caution

Having an unconstrained bit width on the loop exit condition results in control logic larger than needed

Lets first examine the hardware that would be synthesized from the conditional accumulator shown in Example 4-11:

Figure 4-18. Loop With Conditional Bounds

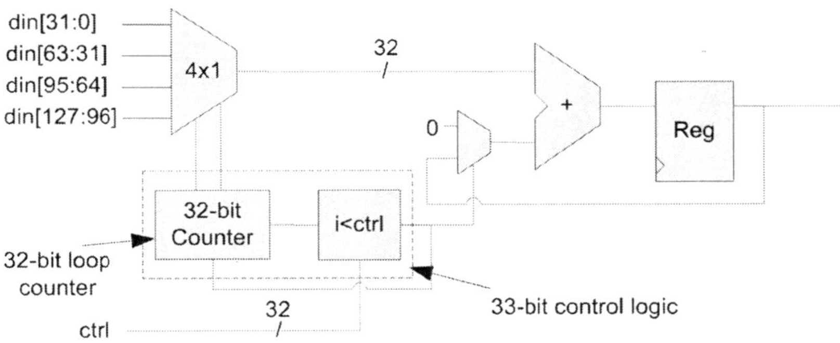

The resulting hardware synthesized from the C++ accumulator with conditional loop bounds results in a 32-bit loop counter and 33-bit logic for the loop exit condition (Figure 4-18). The reason for this is that the interface variable “ctrl” is a 32 bit integer. Because “ctrl” is on the design interface, HLS has no way of knowing, or more importantly proving, that it only should ever range from one to four.

Note

This is an important lesson about HLS in that it only automatically reduces bit-widths where it can symbolically prove that it can be done without changing the functionality between the C++ code and the generated RTL.

In this example, the C++ specifies a 32-bit interface variable which requires 33-bit control logic to be functionally equivalent. The solution to the problem shown above requires two minor C++ code changes, and can be split into two parts; fixing the loop counter, and fixing the loop exit condition.

Optimizing the Loop Counter

In order for HLS to reduce the bit width of the loop counter the loop upper bound should be set to a constant. However, since the execution of each loop iteration is determined by the variable, “ctrl”, we need to add a mechanism for terminating the loop early. This is done by using a conditional break in the loop body shown in Example 4-12.

Example 4-12. Bounded Loop with Conditional Break

```
#include "accum.h"
#include <ac_int.h>
void accumulate(int din[4], int &dout, int ctrl){
  int acc=0;
 ACCUM:for(int i=0;i<4;i++){
    acc += din[i];
    if(i>=ctrl-1)
      break;
  }
  dout = acc;
}
```

The conditional break is placed at the end of the loop because it is assumed that there is at least one loop iteration. Having the conditional break at the end of the loop should give the best quality of results in general. However, if “ctrl” can be zero, meaning that the loop can have zero iterations, the break must be placed at the beginning of the loop body. Figure 4-19 shows the resulting hardware from Example 4-12.

Figure 4-19. Bounded Loop With Conditional Break

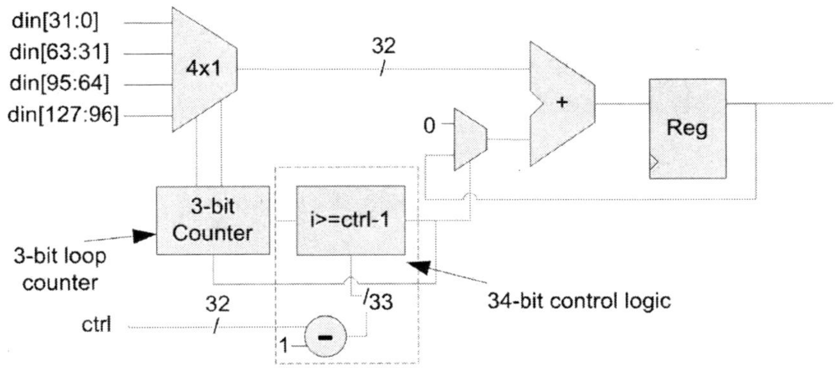

The code transformation has the effect of reducing the loop counter to three bits by fixing the upper loop bound to a constant. Unfortunately, the code change has actually made the design slightly larger. Putting the conditional break at the end of the loop has created a 33-bit subtractor to compute "ctrl-1" and a 34-bit subtractor to compute the ">=" operation. This is in part because "ctrl" is 32-bits and cannot be automatically reduced since it is on the design interface. The control logic can be further optimized.

Note

In general making the loop bounds constant produces better hardware. Conditional breaks can be used inside of the loop to give the same functionality as a variable loop bound.

Optimizing the Loop Control

There are two problems with the control logic for the loop exit condition of Example 4-12:

- Use of a 32-bit integer on the design interface
- Exit condition test requires a subtractor to compare against ctrl-1

HLS is not able to reduce the bit-widths on the top-level design interface for this design since it cannot prove that "ctrl" is always between one and four. In this case the designer must constrain the bit-width of "ctrl" to the desired number of bits. Native C++ data types do not give designers the ability to specify arbitrary bit widths on a variable so bit-accurate data types are required. A better way to write the code to optimize the loop control is shown in Example 4-13.

Example 4-13. Optimized Loop Control

```
1   #include "accum.h"
2   #include <ac_int.h>
3   void accumulate(int din[4], int &dout, ac_int<3,false> ctrl){
4     int acc=0;
5     int i_old=0;
6    ACCUM:for(int i=0;i<4;i++){
7       acc += din[i];
8       if(i_old==ctrl)
9         break;
10      i_old = i;
11    }
12    dout = acc;
13  }
```

The following code changes were made to optimize the loop control logic:

1. Line 3 - "ctrl" was constrained to three bits reducing the comparison logic to three bits
2. Line 10 - "i_old" stores the previous value of the loop iterator "i"
3. Line 8 - The exit condition test is made on the previous value of "i" eliminating the need for a subtractor.

The resulting hardware from Example 4-13 is shown in Figure 4-20.

Figure 4-20. Optimized Loop Control

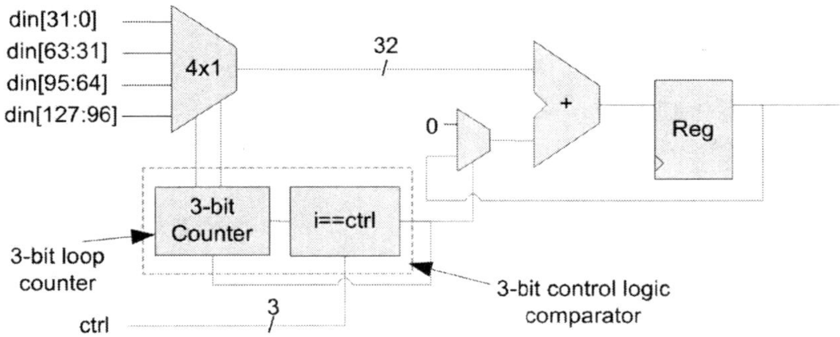

Note

Interface variables should always be constrained to the minimum number of bits, especially when used as a loop control variable.

Nested Loops

Nested loops and the effects of pipelining nested loops is often one of the most misunderstood concepts of high-level C++ synthesis. Understanding the resulting hardware behavior from synthesizing non-pipelined and pipelined nested loops allows designers to more easily meet performance and area requirements. The simple accumulator that has been used in previous examples can be extended to illustrate the effects of nested loops.

Example 4-14. Nested Loop Accumulator

```
#include "accum.h"
#include <ac_int.h>
#define MAX 100000
void accumulate(int din[2][4], int &dout){
  int acc=0;
 ROW:for(int i=0;i<2;i++){
     if(acc>MAX)
       acc = MAX;
  COL:for(int j=0;j<4;j++){
       acc += din[i][j];
     }
  }
  dout = acc;
}
```

The following enhancements to the C++ accumulator designer were made:

1. Line 4- The input data is a 2x4 array of type integer.
2. Lines 6 and 10 - Two loops, ROW and COL are nested to index the rows and columns of the 2x4 array.
3. Lines 7 and 8 - The accumulate variable “acc” is saturated to keep it from exceeding a maximum value at the beginning of the ROW loop. This is somewhat of an artificial example but helps illustrate the effects of nesting loops.

Unconstrained Nested Loops

If the nested loop accumulator in Example 4-14 is synthesized with both loops left rolled and no loop pipelining, the resulting hardware has a behavior similar to the state diagram shown in Figure 4-21. Note that for this example it is assumed that each iteration of the COL loop can execute in one clock cycle.

Figure 4-21. State Diagram of Unconstrained Nested Loops

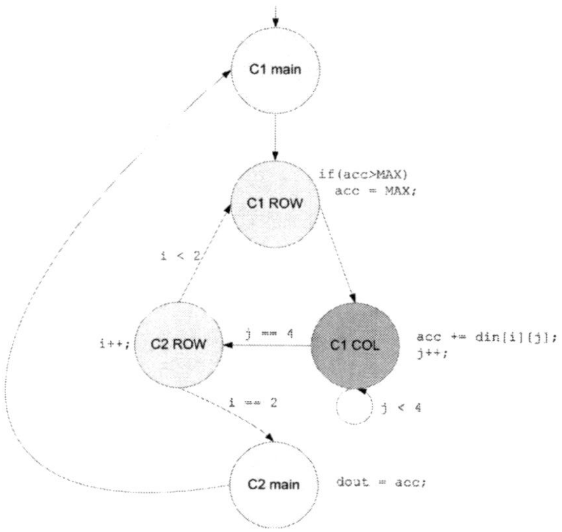

The state diagram in Figure 4-21 shows that unconstrained nested loops have an overhead associated with the computation of the outer loop body and index (C1_ROW and C2_ROW). This overhead has the impact of increasing the latency of the design. This increased latency can be substantial compared to the number of clock cycles required to perform the main computation of the algorithm. Figure 4-21 shows that the execution of the design requires two c-steps (clock cycles) for the main loop, and two c-steps for the ROW loop, in addition to the eight cycles required to execute the COL loop twice. This means that the entire design takes 14 cycles to accumulate the eight values of din[2][4], which in turn means that 43% of the execution time is taken up by the main and ROW loop overhead in this example. Figure 4-22 shows the schedule for the unconstrained design.

Figure 4-22. Schedule of Unconstrained Nested Loops

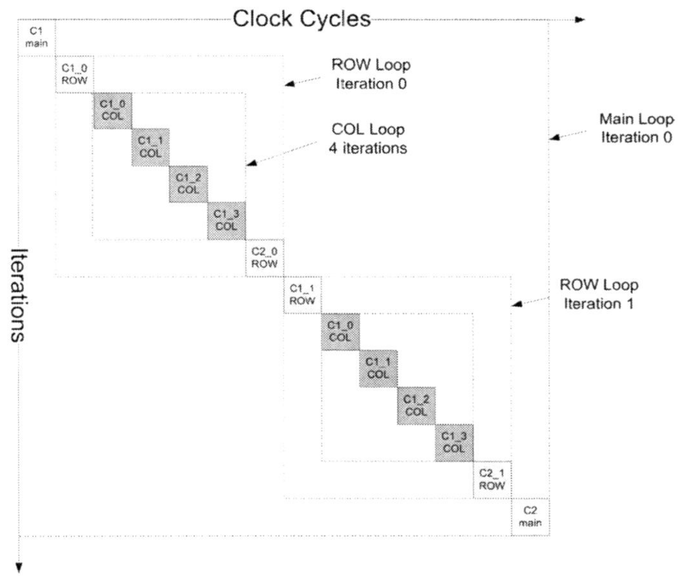

Note

Unconstrained nested loops can increase latency because of the overhead of computing the loop exit conditions and loop bodies separately.

Pipelined Nested Loops

Loop pipelining can be applied in order to improve design performance.

Note

It is generally good practice to begin pipelining starting with the innermost loops and working up towards the top-most loops. This should in general give the best area/performance trade-off.

For this example the innermost loop is the COL loop. However since it was assumed that each iteration of the COL loop only requires one c-step to execute there is no benefit in pipelining this loop.

Pipelined ROW Loop With II=1

Note

When nested loops are pipelined together the loops are "flattened" into a single loop. The initiation interval constraint is then applied to the flattened loop.

Figure 4-23 shows the state diagram that illustrates the effect of pipelining the ROW and COL loops for Example 4-14.

Figure 4-23. State Diagram of ROW Loop with II=1

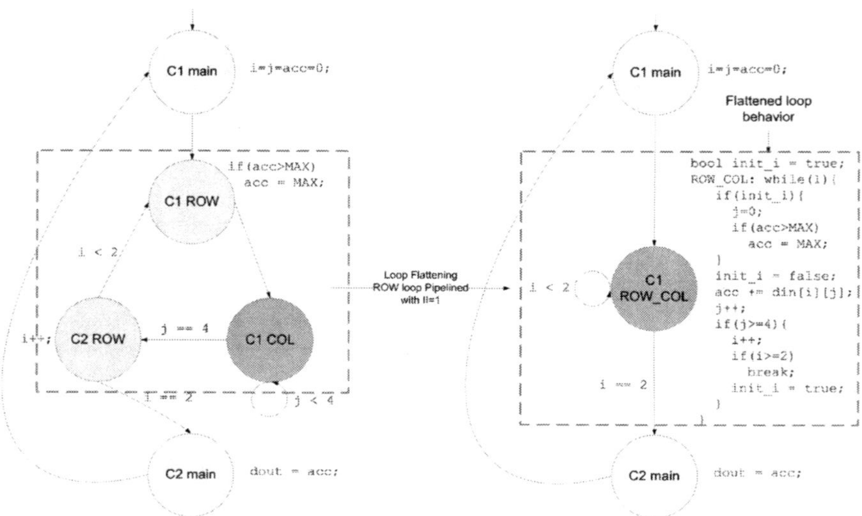

Figure 4-23 shows how loop flattening removes the overhead of the C1_ROW and C2_ROW states by combining the saturation and ROW loop index logic into the same loop with the COL loop. Although pipelining nested loops improves performance in terms of latency and throughput, it is not without cost. The control logic becomes progressively more complex as more and more nested loops are pipelined. This can lead to larger area, or failure to schedule in some cases. So a good rule of thumb is to start pipelining the inner loops and work your way towards the outer loops until the performance target is met. Figure 4-24 shows the schedule when the ROW loop is pipelined with II=1.

Figure 4-24. Schedule of ROW Loop with II=1

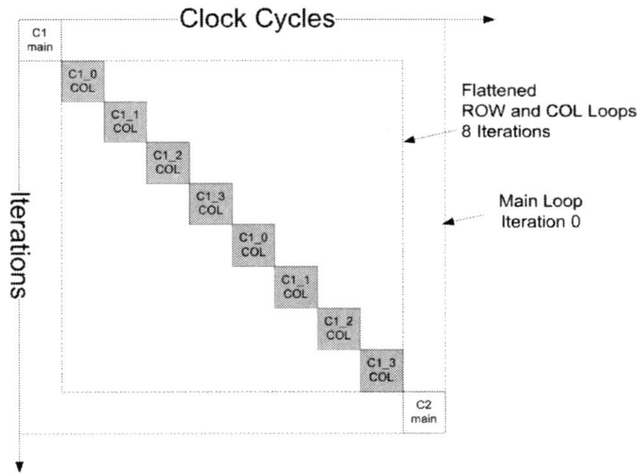

Figure 4-24 shows that by pipelining the ROW and COL loops together, the two cycle overhead of the ROW loop has been absorbed into the flattened loop allowing the nested ROW and COL loops to execute in eight clock cycles. The only overhead remaining is caused by the main loop.

Pipelined main Loop with II=1

Similar to pipelining the ROW loop, pipelining the main loop causes the main, ROW, and COL loops to be flattened into a single loop. This has the effect of moving the loop iterator initialization and the write of the output "dout" into the ROW_COL loop and executing them conditionally, as shown in the state diagram of Figure 4-25. The net result is to increase the design performance at the expense of making the control logic more complicated. Figure 4-26 shows the schedule with the main loop pipelined. The two cycle overhead of the main loop has been flattened along with the ROW and COL loops allowing the design to achieve maximum performance.

Figure 4-25. State Diagram of main Loop with II=1

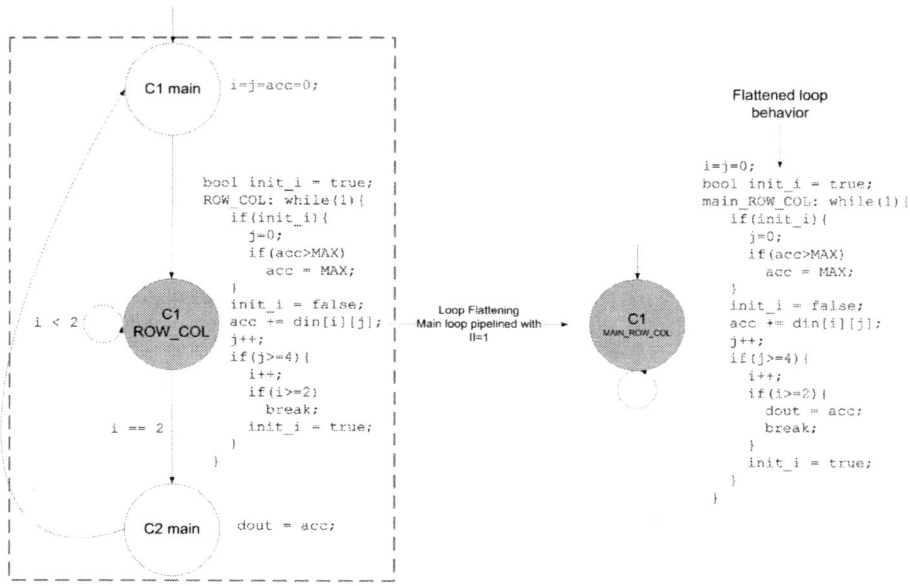

Figure 4-26. Schedule of Main Loop with II=1

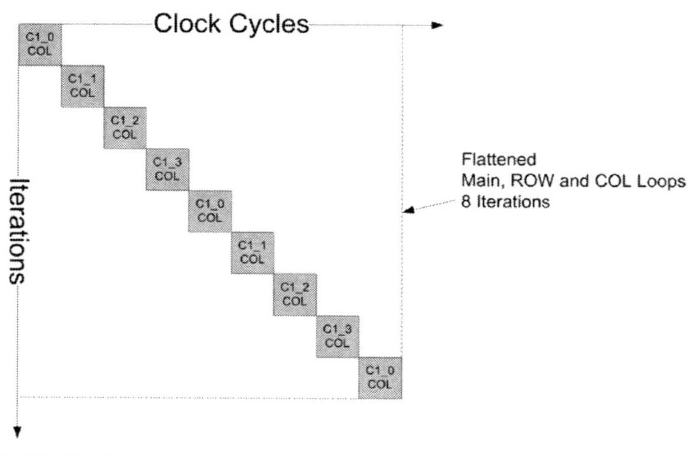

Unrolling Nested Loops

Loop unrolling can be applied to nested loops to increase the design performance, often at the expense of larger area. Because of this, designers must be methodical in choosing how much to unroll a loop. For nested loops with a large number of iterations it is more commonplace to leave the outer loop(s) rolled and partially or fully unroll the inner loop when trying to increase design performance. This is also usually done in combination with loop pipelining.

Note

In general it is always better to pipeline loops first before using loop unrolling. This is because loop pipelining often gives a significant boost in performance with a smaller cost in terms of area.

Loop unrolling on the other hand usually has a greater impact on area when the loop body contains a large number of operations. This is because unrolling replicates the loop body leading to larger numbers of resources being scheduled in parallel.

Unrolling the Innermost Loop

Example 4-15 shows a C++ design that uses two nested loops to separately accumulate the rows of a two-dimensional array. This example is synthesized with the COL loop fully unrolled and the ROW loop pipelined with II=1.

Example 4-15. Unrolling the Inner Loop

```
#include "accum.h"
void accumulate(int din[2][4], int dout[2]){
  int acc[2];
 ROW:for(int i=0;i<2;i++){
     acc[i] = 0;
  COL:for(int j=0;j<4;j++){
      acc[i] += din[i][j];
    }
    dout[i] = acc[i];
  }
}
```

Fully unrolling the COL loop has the same effect as manually replicating the COL loop body, shown in Example 4-16.

Example 4-16. Unrolling the Inner Loop Manually

```
#include "accum.h"
void accumulate(int din[2][4], int dout[2]){
  int acc;
 ROW:for(int i=0;i<2;i++){
    acc=0;
    acc += din[i][0];
    acc += din[i][1];
    acc += din[i][2];
    acc += din[i][3];
    dout[i] = acc;
  }
}
```

Example 4-16, which shows the effects of duplicating the inner loop body is transformed during scheduling into something that more closely resembles the code shown in Example 4-17.

Example 4-17. Unrolling the Inner Loop Transformation

```
#include "accum.h"
void accumulate(int din[2][4], int dout[2]){
  int acc=0;
 ROW:for(int i=0;i<2;i++){
      dout[i] = din[i][0]+din[i][1]+din[i][2]+din[i][3];
    }
  }
}
```

Example 4-17 shows that accumulating four values at a time requires three adders. Assuming that there is sufficient time to schedule the three adders in the same clock cycle, the design schedule looks like that shown in Figure 4-27. Each iteration of the ROW loop executes in one clock cycle, while there is still some overhead caused by not pipelining the main loop.

Figure 4-27. Schedule with Inner Loop Fully Unrolled ROW Loop with II=1

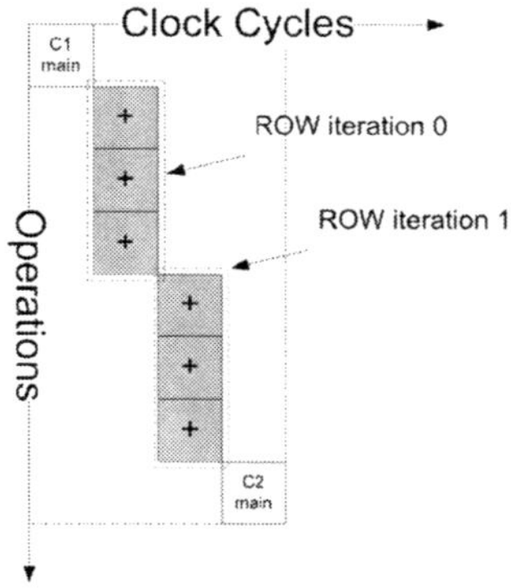

The hardware resulting from synthesizing Example 4-15 is shown in Figure 4-28. High-level synthesis automatically builds a balanced adder tree when unrolling accumulators inside a loop. There are some situations where the tree balancing does not happen automatically when the accumulate is conditional. This is discussed later.

Figure 4-28. Hardware with Inner Loop Fully Unrolled

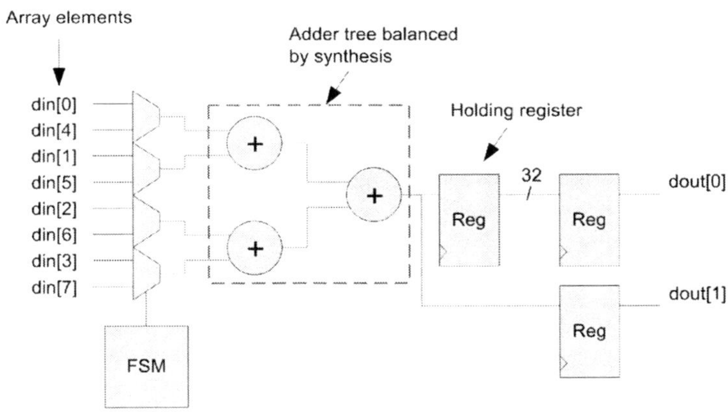

Rampup/Rampdown of Pipelined Nested Loops

Increasing the clock frequency when synthesizing Example 4-15 at some point requires that the adder tree shown in Figure 4-28 be scheduled over multiple clock cycles or c-steps. Figure 4-29 shows the design schedule where the first two adders are scheduled together in the same c-step, with the second adder stage scheduled in the next c-step. Two pipeline stages are created when the ROW loop is pipelined with II=1, and the design latency and throughput is affected due to pipeline rampup and rampdown, initially discussed in "Classic RISC Pipelining" on page 41. For loops with large number of iterations, the effect of rampup/rampdown may be negligible, and allowing the pipeline to rampdown has the added benefit of allowing all data to be "flushed" from the pipeline stages. In this example the cost of rampup/rampdown is significant compared to the number of iterations for the ROW loop.

Figure 4-29. Schedule of Ramp-up/down with Inner Loop Fully Unrolled

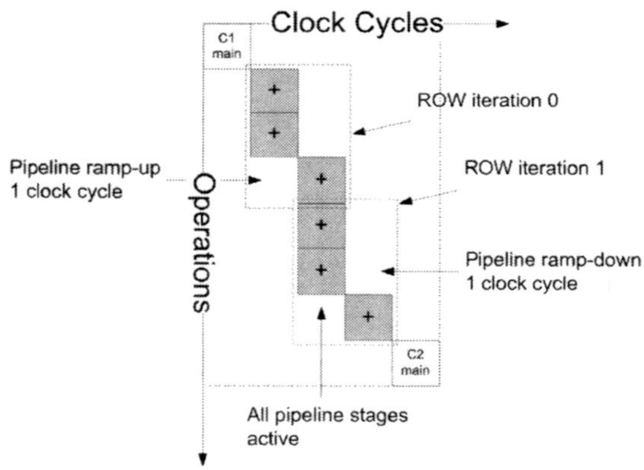

Figure 4-30 shows the hardware generated for the schedule shown in Figure 4-29. The adder tree has been separated into two pipeline stages.

Figure 4-30. Hardware of Ramp-up/down with Inner Loop Fully Unrolled

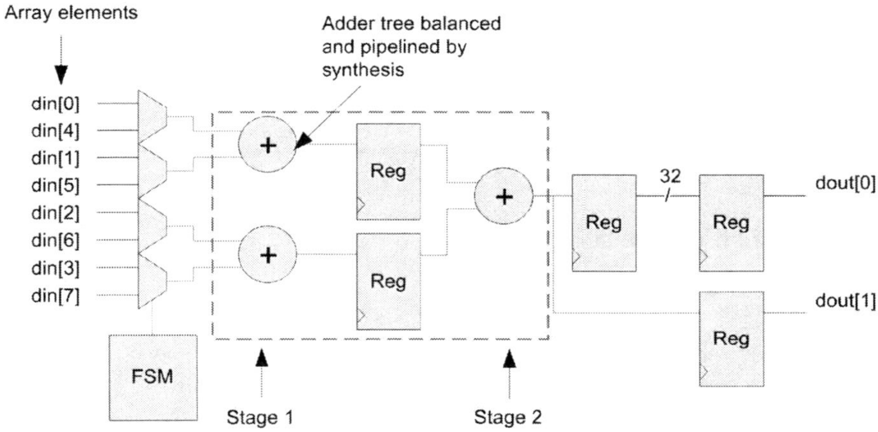

Rampup Only of Nested Loops with Pipelined Main Loop

A possible solution for increasing the performance for designs that have both rampup and rampdown of the pipeline would be to pipeline the "main" loop with II=1. When this is done the pipeline only ramps up and then runs forever, removing the throughput cost of pipeline rampdown. This is shown in Figure 4-31.

Figure 4-31. Rampup of Nested Loops with Main Loop II=1

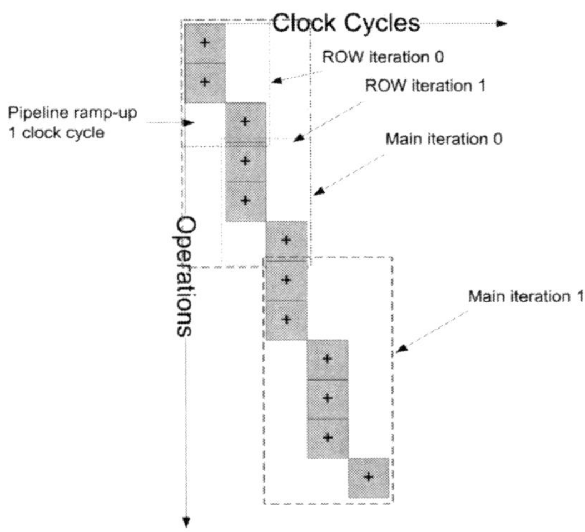

Caution

There are side effects associated with pipelining the main loop when the design has rolled loops. If IO is mapped to a handshaking interface and is accessed inside of the pipelined loop it can cause the pipeline to stall. This is covered in "Conditional IO" on page 90.

Unrolling the Outer Loop

The previous section illustrated how unrolling the innermost loop replicates the loop body resulting in higher performance. The core architectural feature resulting from unrolling the innermost loop was a balanced adder tree, Figure 4-28. If the inner loop is left rolled and the outer loop is unrolled the inner loop is replicated as many times as the loop is unrolled. Example 4-18 shows the effects of manually unrolling the outer loop where there are now two copies of the inner loop, COL_0 and COL_1.

Example 4-18. Manually Unrolling the Outer Loop

```
#include "accum.h"
void accumulate(int din[2][4], int dout[2]){
  int acc[2];

  acc[0] = 0;
 COL_0:for(int j=0;j<4;j++){
    acc[0] += din[0][j];
  }
  dout[0] = acc[0];
  acc[1] = 0;
 COL_1:for(int j=0;j<4;j++){
    acc[1] += din[1][j];
  }
  dout[1] = acc[1];
}
```

When possible, high-level synthesis automatically merges all of the replicated loops into a single loop, leading to a number of accumulators running in parallel. Example 4-19 shows the effects of manually merging the two COL loops.

Example 4-19. Manual Merging

```
#include "accum.h"
void accumulate(int din[2][4], int dout[2]){
  int acc[2];

  acc[0] = 0;
  acc[1] = 0;
 COL_0_1:for(int j=0;j<4;j++){
    acc[0] += din[0][j];
    acc[1] += din[1][j];
  }
  dout[0] = acc[0];
  dout[1] = acc[1];
}
```

Figure 4-32shows the schedule when the ROW is fully unrolled and all copies of the COL loop are merged.

Figure 4-32. Unrolling the Outer Loop with Loop Merging

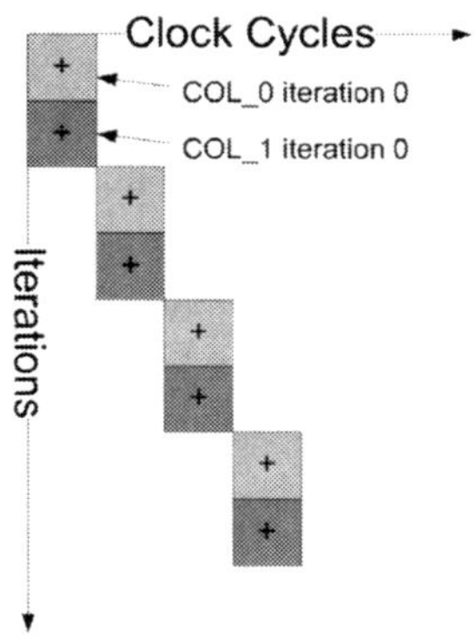

Figure 4-33 shows the synthesized hardware resulting from unrolling the outer loop which has had the effect of creating two accumulators running in parallel.

Figure 4-33. Hardware of Unrolling the Outer Loop

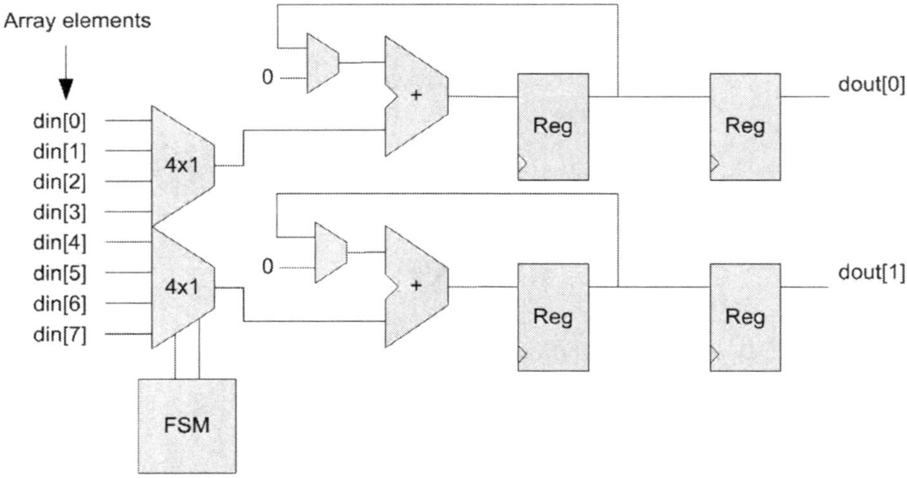

Reversing the Loop Order

The previous section illustrated how unrolling the outer loop cause the inner loop to be replicated and merged automatically during synthesis. However, there are situations that prevent automatic merging, and this leads to sub-optimal performance. Example 4-20 shows the accumulator design used in the previous section that has been modified to conditionally assign the index for the "acc" array. This conditional index assignment breaks automatic loop merging.

Example 4-20. Conditional Index Breaks Loop Merging

```
#include "accum.h"
void accumulate(int din[2][4], int dout[2], bool flag){
  int acc[2];
  int idx;
 ROW:for(int i=0;i<2;i++){
    idx = flag ? i: 1-i;
    acc[idx] = 0;
  COL:for(int j=0;j<4;j++){
      acc[idx] += din[i][j];
    }
    dout[i] = acc[i];
  }
}
```

Not merging the two copies of the COL loop that result from unrolling the ROW loop causes the loops to be scheduled sequentially (See

"Sequential Loops" on page 69). Pipelining the main loop with II=1 causes the two copies of the COL loop, COL_0 and COL_1, to be flattened into the main loop, but they are still be executed sequentially as shown in Figure 4-34.

Figure 4-34. Schedule with Conditional Index and ROW Loop Unrolled

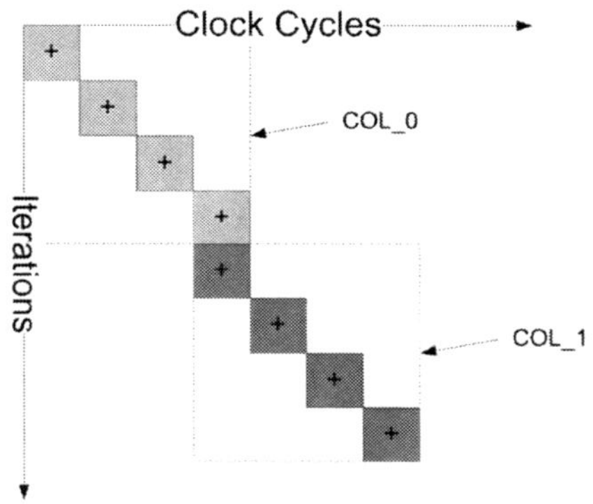

One possible solution to achieve the desired behavior of two accumulators running in parallel is to reverse the order of the ROW and COL loops. However, this must be done carefully since it usually requires that the outer loop body must be moved to the inner loop and executed conditionally. Example 4-21 shows how to manually reverse the loop order.

Example 4-21. Reversing the Loop Order

```
#include "accum.h"
void accumulate(int din[2][4], int dout[2], bool flag){
  int acc[2];
  int idx;

 COL:for(int j=0;j<4;j++){
  ROW:for(int i=0;i<2;i++){
      idx = flag ? i: 1-i;
      if(j==0)
        acc[idx] = 0;
      acc[idx] += din[i][j];
      if(j==3)
        dout[i] = acc[i];
    }
  }
}
```

The following code changes were made in Example 4-21.

- Lines 6 and 7- reversed the order of the ROW and COL loops
- Line 8- Moved the index computation into the inner loop body
- Lines 9 and 10 - Moved the clearing of the accumulators into the inner loop body and made it conditional so that they are only cleared once at the beginning
- Lines 12 and 13 - Moved the writing of the output into the inner loop body and made the writes conditional so that the output is only written on the final iteration of COL

Sequential Loops

It is not uncommon to have multiple consecutive loops in a C++ design. Although these loops execute sequentially in the simulation of the C++, HLS can be directed to automatically merge these loops and execute them in parallel in hardware. However there are many cases where the C++ code can be written in such a way as to make automatic loop merging impossible. In these cases either the C++ code must be re-written to manually merge the loops if better performance is required, or explicit hierarchy should be used (See "Hierarchical Design" on page 191).

It is important for designers to understand the behavior of the hardware when loop merging does and does not take place so there are no unexpected results.

Simple Independent Sequential Loops

Example 4-22 shows the case where there are two sequential loops that are used to separately accumulate two four-element arrays.

Example 4-22. Independent Sequential Loops

```
#include "accum.h"
void accumulate(int din0[4], int din1[4],int &dout0, int &dout1)
{
  int acc0=0;
  int acc1=0;
 ACCUM0:for(int i=0;i<4;i++){
    acc0 += din0[i];
  }
```

```
    ACCUM1:for(int i=0;i<4;i++){
       acc1 += din1[i];
     }
     dout0 = acc0;
     dout1 = acc1;
   }
```

High-level synthesis can automatically merge these loops because there are no dependencies between the loops and the indexing of the arrays is based solely on the loop iterators. With loops left rolled and automatically merged, and the main loop pipelined with II=1, the resulting schedule looks like that shown in Figure 4-35

Figure 4-35. Schedule of Merged Sequential Loops

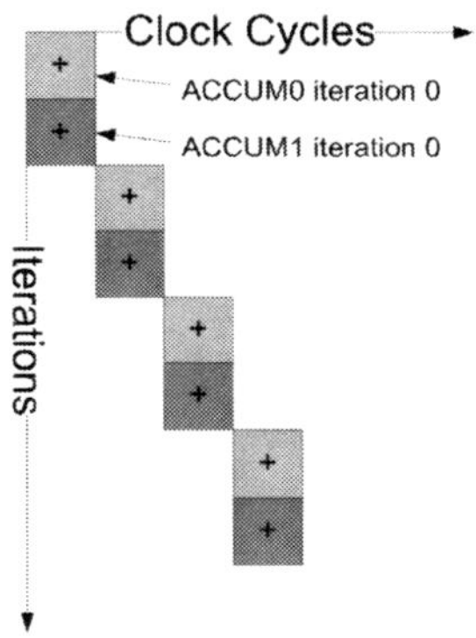

The schedule shown above indicates that the loop iterations in each of the ACCUM loops can be run at the same time, resulting in a design that has two accumulators and runs in four clock cycles (Figure 4-36). If this kind of performance and increase in area is not required, automatic loop merging can be disabled during synthesis, allowing the loops to execute sequentially. This is discussed in the next section.

Figure 4-36. Hardware of Merged Sequential Loops

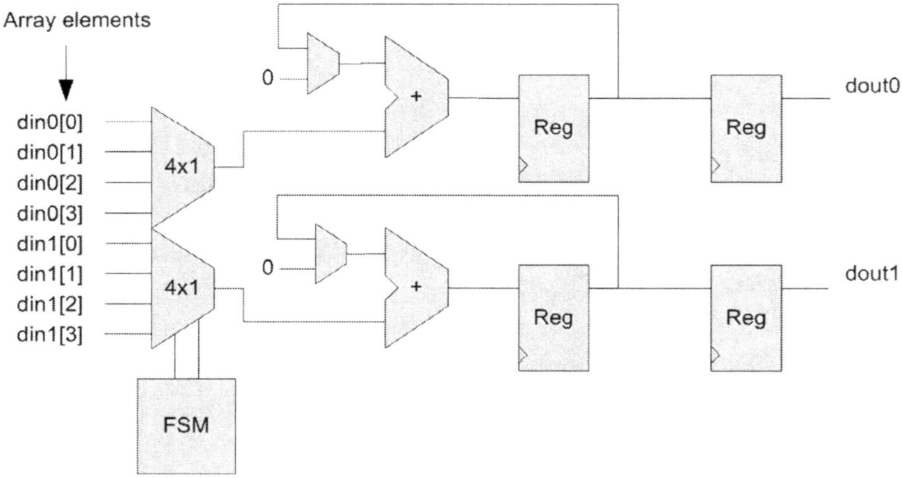

Effects of Unmerged Sequential Loops

In some instances sequential loops are not automatically merged. This can occur either intentionally because the design does not require the extra performance, usually at the cost of higher area, or because there are dependencies between the loops that break loop merging optimizations. Other operations such as conditional index assignment for reading or writing an array can also prevent loop merging optimizations. In either of these cases it results in designs that have both longer latency and throughput.

Consider the following design shown in Example 4-23. In this example the accumulated result from the ACCUM0 loop is used as the starting value for the ACCUM1 loop. These loops are not automatically merged since the ACCUM0 loop must finish before the ACCUM1 loop can start.

Example 4-23. Unmerged Sequential Loops

```
#include "accum.h"
void accumulate(int din0[4], int din1[4],int &dout0, int &dout1)
{
  int acc0=0;
  int acc1=0;

 ACCUM0:for(int i=0;i<4;i++){
    acc0 += din0[i];
  }
  acc1 = acc0;
```

```
  ACCUM1:for(int i=0;i<4;i++){
     acc1 += din1[i];
   }
   dout0 = acc0;
   dout1 = acc1;
}
```

Figure 4-37 shows the schedule when the main loop of Example 4-23 is pipelined with an II=1. It also illustrates the effect of pipelining the main loop when there are unmerged sequential loops in the design. Pipelining the main loop causes all loops in the design to be flattened, which in turn causes the last iteration of the ACCUM0 loop to be overlapped with the first iteration of the ACCUM1 loop. Although this improves the design performance slightly it has the impact of requiring two adders to implement the hardware. If performance is not an issue it is better to pipeline the ACCUM0 and ACCUM1 loops individually. This should then allow the operations scheduled in each loop to be shared, reducing the area. However pipelining the loops individually can impact the performance since each loop must then ramp-up and ramp-down separately.

Figure 4-37. Schedule of Unmerged Sequential Loops with Main II=1

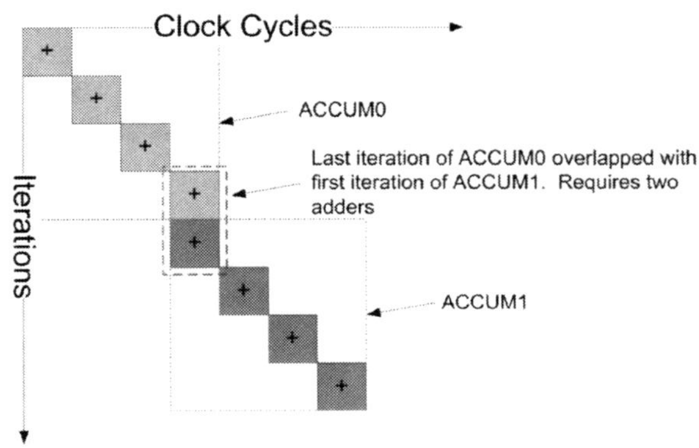

Figure 4-38 shows the schedule when the ACCUM0 and ACCUM1 loops of Example 4-23 are pipelined with II=1 instead of pipelining the main loop. In this case there is no overlap between the loops and a single adder can be used to implement the hardware. However there is a two cycle performance penalty incurred due to the un-pipelined main loop (C1 Main and C2 Main).

Figure 4-38. Schedule of Unmerged Sequential Loops with ACCUM(s) II=1

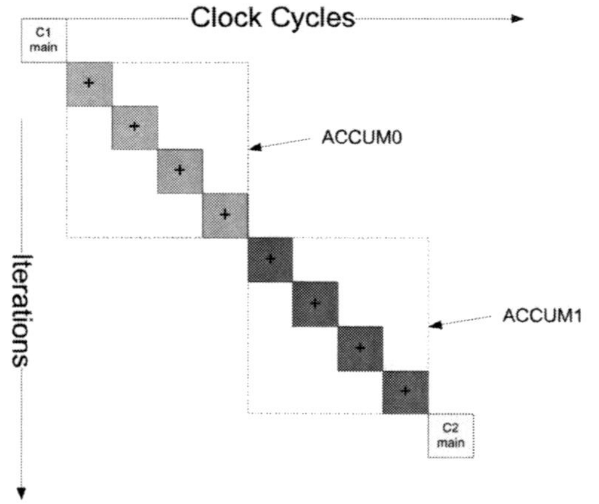

Manual merging of sequential loops

It is up to the designer to manually merge sequential loops in situations where HLS does not do it automatically, and merged loops is the desired behavior. This usually means rewriting the C++ code. Example 4-24 shows the manual rewrite of the code in Example 4-23 in order to achieve the best possible performance.

Example 4-24. Manually Merged Sequential Loops with Main II=1

```
#include "accum.h"
void accumulate(int din0[4], int din1[4],int &dout0, int &dout1)
{
  int acc0=0;
  int acc1=0;
  int tmp;
 ACCUM0_1:for(int i=0;i<4;i++){
    tmp = din0[i];
    acc0 += tmp;
    acc1 += din1[i]+tmp;
  }
  dout0 = acc0;
  dout1 = acc1;
}
```

The example shown above manually merged the sequential loops so that the design runs in four clock cycles when pipelining the main loop with II=1. However this design is larger than the previous implementations because it requires three adders, shown in the schedule in Figure 4-39.

Figure 4-39. Schedule of Manual Merged Sequential Loops with Main II=1

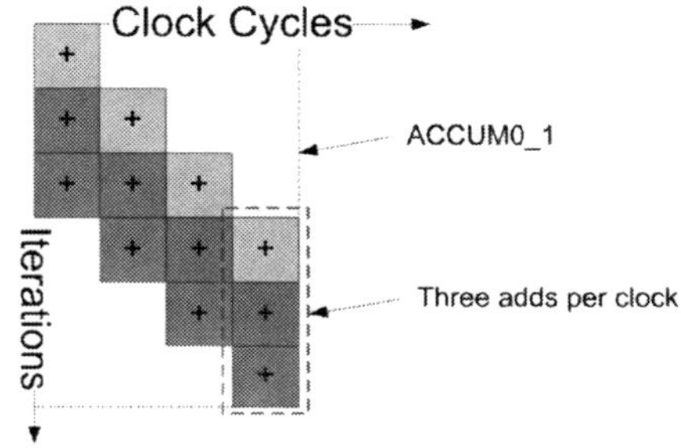

Pipeline Feedback

The initiation interval can be set anywhere from a synthesis tool dependent maximum down to an II=1 on any feed-forward design. However, a design with feedback limits the initiation interval to be no less than the delay of the feedback path. There are three types of feedback, data dependent, control dependent, and inter-block feedback. Inter-block feedback is discussed in later chapters covering system level design.

Data Feedback

Data feedback occurs when the input to a data path operation is dependent on a variable computed in the previous loop iteration. If the only loop in the design is the main loop the variable must have been declared as static for there to be feedback. Consider the following design:

Example 4-25. Data Feedback Design

```
void accumulate(int a, int b, int &dout){
  static int acc=0;
  int  tmp = acc*a;
  acc = tmp+b;
  dout = acc;
}
```

```
Design Constraints
Clock frequency slow
Main loop pipelined with II=1
```

If the clock frequency for this design is assumed to be very slow the schedule and hardware would look approximately like Figures 4-40 and Figure 4-41. The design schedule shows that pipelining with II=1 is possible since each iteration of the main loop finishes computing "acc" before the next iteration starts. This is also obvious by looking at the hardware diagram.

Figure 4-40. Feedback Within One Clock Cycle

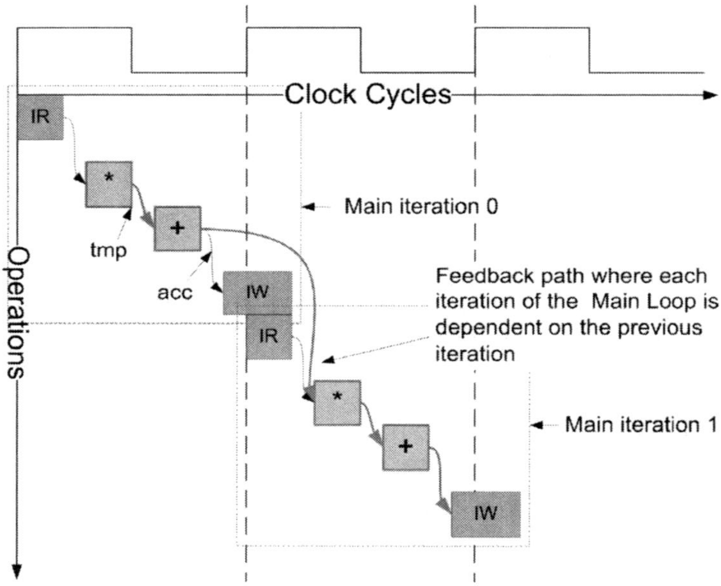

Figure 4-41. Hardware for Feedback Within One Clock Cycle

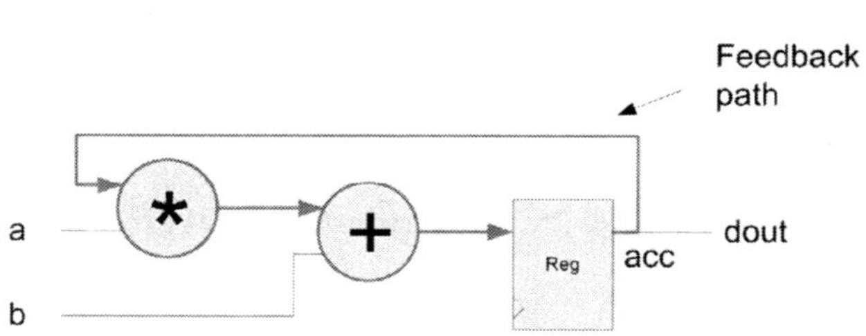

Now consider the same design from Example 4-25 re-synthesized with the following constraints:

```
Design Constraints
Clock frequency very fast
```

```
Main loop pipelined with II=1
Multiplier constrained to a two-cycle pipelined multiplier
```

This design cannot be pipelined with II=1 with the given set of constraints listed above. The failed schedule shown in Figure 4-42 illustrates why. To pipeline with II=1 would mean that “acc” is available to be read in the second clock cycle. However, the first pipeline stage is not finished computing “acc” until the edge of the third clock cycle. Another way to look at this is to examine the hardware that is synthesized, shown in Figure 4-43. It takes two clock cycles to compute “tmp” in the feed-forward path. “tmp” is then added to the current value of “b” and fed back to the multiplier. Lines 3 and 4 of Example 4-25 show that each time a new value of “acc” is computed it is available in the next iteration to compute “tmp”. Thus the hardware pipeline cannot be made to run every clock cycle since it must allow the multiplier to flush for each computation of “tmp”. The best possible performance would be pipelining with II=2.

Figure 4-42. Failed Schedule for Multi-cycle Feedback

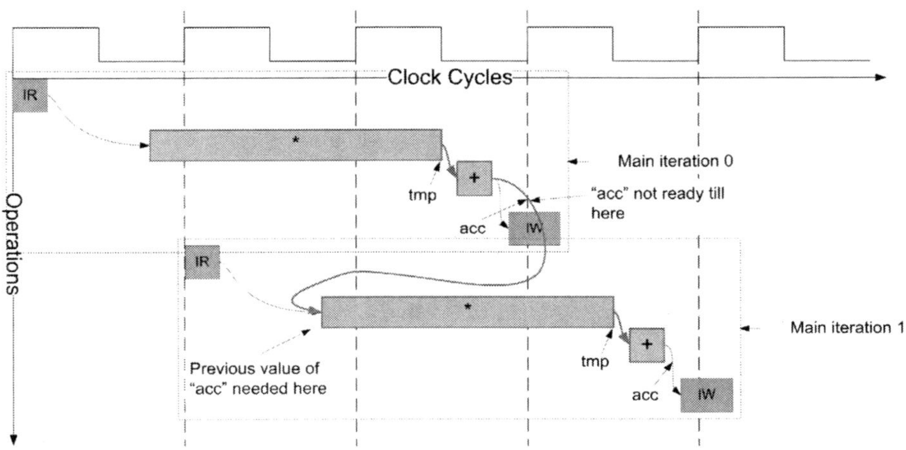

Figure 4-43. Hardware for Multi-cycle Feedback

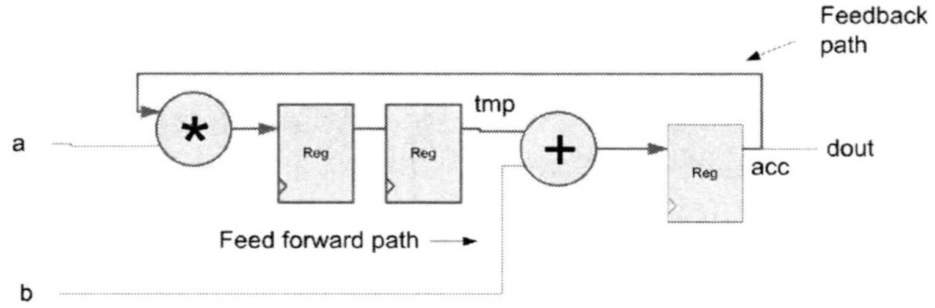

The solution to getting the design discussed above to pipeline with II=1 is to modify the design to balance the delays between the feed-forward and feedback paths. This means introducing delay elements in the C++ along the feedback path. The functionality is different from the original design, but there is no other way to pipeline with II=1 and have the RTL match the C++ exactly. Example 4-26 shows Example 4-25 rewritten to balance the delay along the feedback path to match the two cycle feed forward delay. This is done by creating a two element shift register to delay “acc”. The hardware synthesized for Example 4-26 is shown in Figure 4-44.

In general the number of shift register elements needed in the feedback path can be computed as:

```
Num Shift Elements = (feed-forward latency)/Initiation Interval
(II)
```

Example 4-26. Balancing Feedback Path Delays

```
1   void accumulate(int a, int b, int &dout){
2     static int acc=0;
3     static int acc_old0;
4     static int acc_old1;
5
6     int  tmp0 = acc_old1*a;
7     acc = tmp0+b;
8     acc_old1 = acc_old0;
9     acc_old0 = acc;
10    dout = acc;
11  }
```

The details of Example 4-26 are:

- Lines 3 and 4 define two static variables used to implement the feedback delays.

- Line 6 uses the delayed feedback "acc_old1" as the input to the multiplier.
- Lines 8 and 9 implement the shift register to delay "acc" by two clock cycles.

Figure 4-44. Hardware with Balanced Delays on Feedback Path

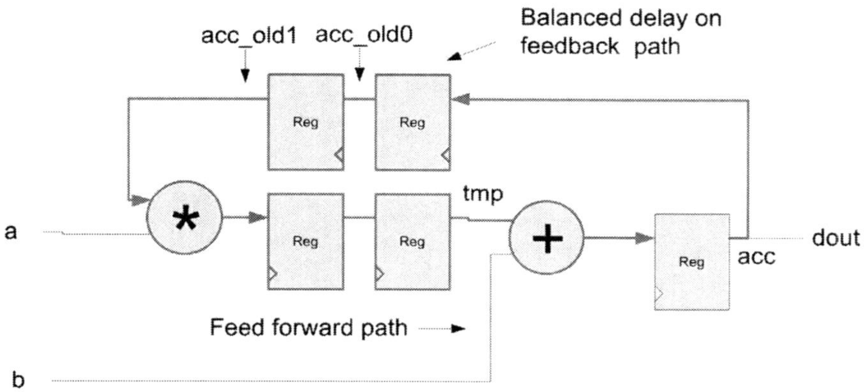

Control Feedback

Pipelining failures due to feedback are also possible due to the loop control in a design. The deeper the nesting of loops in a design, the more complicated the control becomes, which in turn limits the clock frequency and ability to pipeline a design. Adhering to the recommended coding practices eliminates many of these potential issues. The following design, Example 4-27, illustrates how "bad" coding style can lead to problems when trying to pipeline. This design does not only have larger area than needed, but also fails pipelining for high clock frequencies due to control feedback. The cause of this is due to the 32-bit interface variables being used for the loop upper bounds. The impact of writing the C++ this way was covered in detail in "Optimizing the Loop Counter" on page 54 and "Optimizing the Loop Control" on page 55. Essentially there is a long combinational path created to evaluate the loop exit conditions. The outer loop "X" has to know when the inner loops are finished so it can exit immediately. Figure 4-45 shows the approximate hardware structure for Example 4-27. Although this is a very rough approximation it clearly shows that there is a combinational path through both 32-bit loop bounds comparisons, which severely impacts performance as the clock frequency is increased. A secondary problem is that the unbounded loops generate 32-bit logic for the loop counters. This can also prevent pipelining due to the feedback on the loop accumulator.

Example 4-27. Control Feedback

```
void control(int din[8][8],
             int dout[8],
             int x_size,
             int y_size){
  int acc;
 X:for(int x=0;x<x_size;x++){
    acc = 0;
  Y:for(int y=0;y<y_size;y++){
      acc += din[x][y];
      dout[x] = acc;
    }
  }
}
```

Figure 4-45. Control Feedback

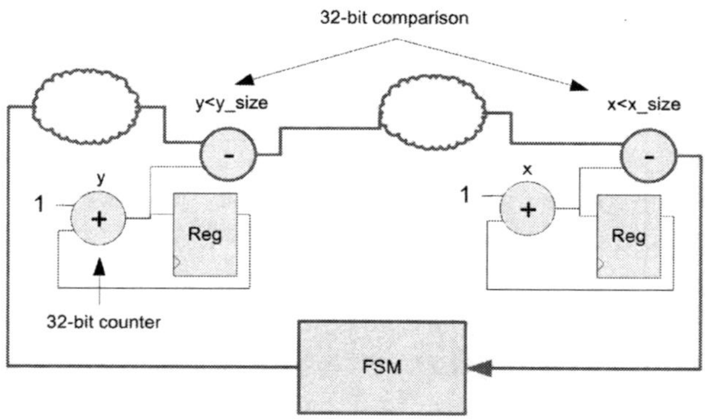

To minimize the possibility of feedback failures, Example 4-27 should be rewritten using the recommended style discussed previously. This is shown below in Example 4-28. The loops have been bounded, and the control logic for the loop exit reduced by using the appropriate bit widths on "x_size" and "y_size"

Example 4-28. Minimizing Control Feedback

```
#include <ac_int.h>
void control(int din[8][8],
             int dout[8],
             ac_int<4,false> x_size,
```

```
                  ac_int<4,false> y_size){
  int acc;
 X:for(int x=0;x<8;x++){
    acc = 0;
  Y:for(int y=0;y<8;y++){
      acc += din[x][y];
      dout[x] = acc;
      if(y==y_size-1)
        break;
    }
    if(x==x_size-1)
      break;
  }
}
```

Conditions

Sharing

HLS can automatically share resources when it can prove mutual exclusivity. This means that HLS can theoretically share any similar operators that are in mutually exclusive branches of a condition, no matter how deeply nested the condition. The reality is that there are a number of ways that the C++ can be written and/or constrained so that the proof of mutual exclusivity is not possible. Usually this is due to a combination of either bad coding style or overly complex or deeply nested conditions. Good coding practices should always allow the maximum amount of sharing.

Conditional expressions are specified using the switch-case and if-else statements.

if-else statement

The if-else statement has the following form:

```
if( condition0 ) {
      statement-list0;
}
else if( condition1 ) {
      statement-list1;
}
...
else {
      statement-listN;
}
```

The conditions evaluate to a boolean expression and can range from simple boolean conditions to complex function calls. The statement list can be any number of C++ assignments, conditional expressions, or function calls.

switch statement

The switch statement has the following form:

```
switch( expression ) {
    case 0: statement list0;
    break;
    case 1: statement list1;
    break;
    ...
    case N: statement listN;
    break;
    default: statement list;
    break;
}
```

The "expression" is typically an integer that selects one of the possible cases. The statement list can be any number of C++ assignments, conditional expressions, or function calls. The statement list for a selected case executes and is followed by a break.

Note

Although it is possible to have a "case" without a "break" this is not generally good for synthesizable C++. The behavior in C++ is to drop through to the next "case". However in C++ synthesis this can sometimes cause replication of logic.

Keep it Simple

Think about what you want the hardware to do and code your design using good design practices. While it is easy to write complex deeply nested conditions and rely on the HLS tool to share everything, it is just as likely to get less sharing than expected. Consider the following design example (Example 4-29) that conditionally accumulates one of four different arrays based on several IO variables. Each condition branch calls the "acc" function with one of four arrays as the input.

Example 4-29. Automatic Sharing and Nested Conditions

```
int acc(int data[4]){
  int tmp = 0;
  ACC:for(int i=0;i<4;i++)
    tmp += data[i];
  return tmp;
}
void test(int a[4], int b[4], int c[4], int d[4],
          bool sel0, bool sel1, bool sel2, int &dout){
  int tmp;
  if(sel0){
    if(sel1)
      tmp = acc(a);
    else if(sel2)
      tmp = acc(b);
    else
      tmp = acc(c);
  }else
    tmp = acc(d);
  dout = tmp;
}
```

```
Design Constraints
Clock frequency slow
Main loop pipelined with II=1
All loops unrolled
```

There are several potential problems with the design in Example 4-29.

1. The four calls to the "acc" function are by default all inlined during synthesis. This means that there are four copies of the "ACC" loop that are inlined and optimized. Although it is possible that HLS can still share everything this will in general lead to longer synthesis runtimes since all four copies must be merged back together and shared. One possible solution to improve sharing and runtime would be to make "acc" into a component using a HLS component flow.
2. Even if everything is shared it is likely that HLS will perform fine-grained sharing, which leads to more MUX logic since each individual operator is shared separately. One possible solution to minimize MUX logic would be to make "acc" into a component.
3. The conditions in this example are simple and the clock frequency is slow enough so that everything is scheduled in the same clock cycle. As the conditions become more complex, the nesting becomes deeper, and/or the clock frequency increases, it is likely that operators will be scheduled in different clock cycles. This can limit sharing. Making "acc" into a component will not help in

these types of situations. The best solution is to rewrite the code so that "acc" is called once.

Example 4-30 shows Example 4-29 rewritten to facilitate sharing. The key is to use the conditions to compute the MUXing of data and control and call the function only once.

Example 4-30. Explicit Sharing and Nested Conditions

```
int acc(int data[4]){
  int tmp = 0;
  ACC:for(int i=0;i<4;i++)
    tmp += data[i];
  return tmp;
}
void test(int a[4], int b[4], int c[4], int d[4],
          bool sel0, bool sel1, bool sel2, int &dout){
  int tmp,data[4];
  for(int i=0;i<4;i++)
    if(sel0){
      if(sel1)
        data[i] = a[i];
      else if(sel2)
        data[i] = b[i];
      else
        data[i] = c[i];
    }else
      data[i] = d[i];
  tmp = acc(data);
  dout = tmp;
}
```

```
Design Constraints
Clock frequency slow
Main loop pipelined with II=1
All loops unrolled
```

Example 4-30 will in general give better results than Example 4-29. This is because the nested conditional expression is only used to control the selection of the input array. Once the input array is selected the "acc" function is called once. Doing this allows HLS to easily optimize the adder tree for the "acc" function and the MUX logic is only needed to select the input data. In essence this is coarse grained sharing. This style should be used when using component flows does not give the desired sharing. This example will also have better runtime in general since the "ACC" loop is only inlined once.

Functions and Multiple Conditional Returns

Although multiple returns in function calls are allowed by both C++ and HLS, they are in general a bad idea. This is true both from a code debugging perspective as well as a synthesis quality of results issue. HLS balances the pipeline stages of all conditional branches. Having a return in the branch complicates this and makes it more difficult to pipeline a design. It is best to use a single return at the end of the function. It's especially bad to use multiple returns to try and make things mutually exclusive.

Consider the following example:

Example 4-31. Multiple Conditional Returns

```
1   int acc(int data[4]){
2     int tmp = 0;
3    ACC:for(int i=0;i<4;i++)
4       tmp += data[i];
5     return tmp;
6   }
7   int test(int a[4], int b[4], bool sel0, bool sel1){
8     if(sel0){
9       return acc(a);
10    }
11    if(sel1){
12      return acc(b);
13    }
14  }
```

Example 4-31 has several problems that will prevent good QofR.

1. Although sel0 and sel1 are mutually exclusive in that they are never evaluated together, HLS will typically not be able to prove this and will not share "acc".
2. The function returns on both lines 9 and 12. If HLS cannot prove mutual exclusivity it will not be able to pipeline with II=1 if the function return is mapped to an IO.
3. The function only returns if "sel0" or "sel1" is true. This means that the return value can be undefined. This undefined behavior may cause logic to be optimized away or simulation behavior of the RTL may not match the C++.

Example 4-31 is rewritten to have only one return, shown below.

Example 4-32. Single Function Return

```
1   int acc(int data[4]){
2     int tmp = 0;
3    ACC:for(int i=0;i<4;i++)
4       tmp += data[i];
5     return tmp;
6   }
7   int test(int a[4], int b[4], bool sel0, bool sel1){
8     int tmp = 0;
9     if(sel0){
10      tmp = acc(a);
11    }
12    else if(sel1){
13      tmp = acc(b);
14    }
15    return tmp;
16  }
```

Example 4-32 has made the conditions mutually exclusive. A temporary variable “tmp” is used to store the result of each condition and is then returned on line 15. The temporary variable is initialized to zero as well so the return value will never be undefined.

Replacing Conditional Returns with Flags

It is also possible to bypass entire sections of code by using a conditional return. This should also be avoided. It is always possible to replace the conditional return with a flag variable that can bypass the code. Consider the following code fragment:

Example 4-33. Conditional Return to Bypass Code

```
...
tmp = 0;
tmp += a;
if(sel0)
   return tmp;
tmp += b;
return tmp;
```

Example 4-33 only adds “b” to “tmp” if “sel0” is false. It should be rewritten as:

Example 4-34. Using Flags to Bypass Code

```
...
bool flagl = false;
tmp = 0;
tmp += a;
if(sel0)
    flag = true;
if(flag)
    tmp += b;
return tmp;
```

Example 4-34 replaces the conditional return with a flag that is set conditionally. The flag is then used to conditionally bypass the same sections of code that were bypassed by the conditional return. A single return is used at the end of the function.

Note

A function should have one and only one return.

References

1. John P. Elliot - Understanding Behavioral Synthesis, Kluwer Academic Publishers 1999

Chapter 5

Scheduling of IO and Memories

Introduction

Similar to loop pipelining and loop unrolling, the way in which IO and memory accesses are coded in a design can have a significant impact on both area and performance. IO and memory accesses tend to be the bottleneck in a system and they can potentially limit the ability to pipeline a design, or negate the benefits gained from loop unrolling. In the worst case using bad style when coding IO or memories prevents scheduling a design.

There are two primary ways for passing IO into and out of a design, pass by value and pass by pointer or reference, which includes arrays. Using one over the other can lead to very different behavior.

Unconditional IO

Unconditional IO is considered to be an interface variable mapped to a "wire" type resource. In other words there is no handshaking protocol and it is assumed that the IO can be accessed at any point in the schedule. This has several ramifications in terms of what hardware is built, as well as how the IO should be dealt with external to the design.

Note

If an IO is unconditional HLS is free to move the IO into and out of conditions, as well as into different c-steps, in order to reduce register area.

Because the IO is a wire type interface there is no signaling mechanism to the external world that indicates when the IO access occurs. It is the responsibility of the designer to ensure that the IO data is set up and available for reading, or ready to accept writing of data. Unconditional IO is used most often for either control type interfaces, where the IO does not change, on in designs that are pipelined with II=1 and the IO is read or written every clock cycle. Otherwise a designer must look at the design schedule in order to determine the correct point in time when IO is accessed. The understanding of this type of IO behavior can be further complicated when either passing by value or passing by reference. The following sections look at each of these cases individually.

Pass by Reference

Pass by reference is when a variable is declared as either a pointer or a reference on the design interface. This means that the data that the variable "points to" or "refers to" is stored externally. In other words the "data" is stored off-chip. What this can often imply when hardware is synthesized is that either the data is expected every clock cycle, or there is some sort of off-chip storage in either registers or memory.

Consider Example 5-1 where the four element accumulator has been enhanced to saturate to a maximum value when a control flag is set to true.

Example 5-1. Unconditional IO Passed by Reference

```
void accumulate(int din[4], int &dout, int &threshold, bool &flag)
{
  int acc=0;
 ACCUM:for(int i=0;i<4;i++){
    acc += din[i];
    if(flag)
      if(acc > threshold)
        acc = threshold;
  }
  dout = acc;
}

Design constraints
ACCUM loop with II=1
All IO mapped to wire interfaces
```

Figure 5-1 shows the schedule for Example 5-1. What this shows is that din[], flag, and threshold are read for each iteration of the ACCUM loop. Because the IO is unconditional the designer must ensure that the data is setup and held for the duration of the ACCUM loop, and the write of "dout" must be captured in C2_Main. Without explicit handshaking this would require counting clock cycles after the design is reset.

Figure 5-1. Schedule of Unconditional IO Passed by Reference

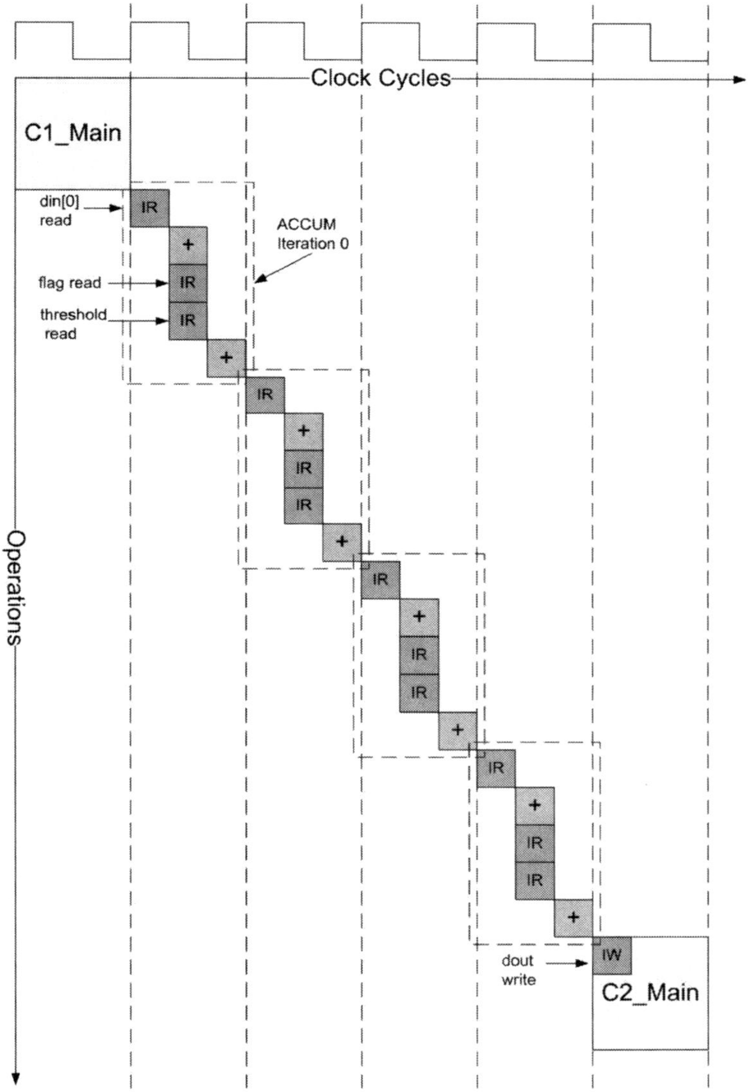

The resulting hardware and typical off-chip configuration of Example 5-1 is shown in Figure 5-2.

Figure 5-2. Hardware of Unconditional IO Passed by Reference

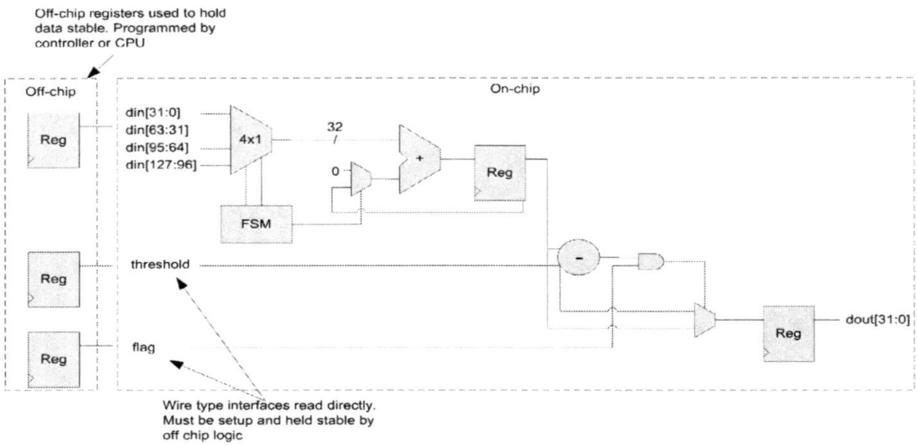

One of the potential problems with the design shown in Figure 5-2 is that there is no synchronization to indicate when din, threshold, and flag are read, or where "dout" is written. This means that the designer must design the external logic to guarantee that the IO data is available at the right point in time. In many cases it is better to add explicit synchronization on the IO. This is covered in later sections.

Pass by Value

Declaring a variable on the design interface as "pass by value" has the effect of registering the data internally in the design. The reason for this is because it matches the behavior of C++ compilers, which is to push pass-by-value interface variables onto the stack when a function is called. The function then pops these variables of the stack and can use them internally, but cannot modify them. The advantage of using pass-by-value versus pass-by-reference is twofold;

1) The IO data does not have to be held stable after it is read at the beginning of the main loop
2) IO traffic is reduced because the data is read once.

Example 5-2 is functionally the same as Example 5-1 but the interface variables "threshold" and "flag" have been made pass-by-value.

Example 5-2. Unconditional IO Passed by Value

```
void accumulate(int din[4], int &dout, int threshold, bool flag)
{
  int acc=0;
 ACCUM:for(int i=0;i<4;i++){
    acc += din[i];
    if(flag)
      if(acc > threshold)
        acc = threshold;
  }
  dout = acc;
}
```

The schedule for Example 5-2 is shown in Figure 5-3. Making the threshold and flag variables pass-by-reference has the effect of reading them once and storing the data in registers at the beginning of the design, eliminating the need to read them for each iteration of the ACCUM loop.

Figure 5-3. Schedule of Unconditional IO Passed by Value

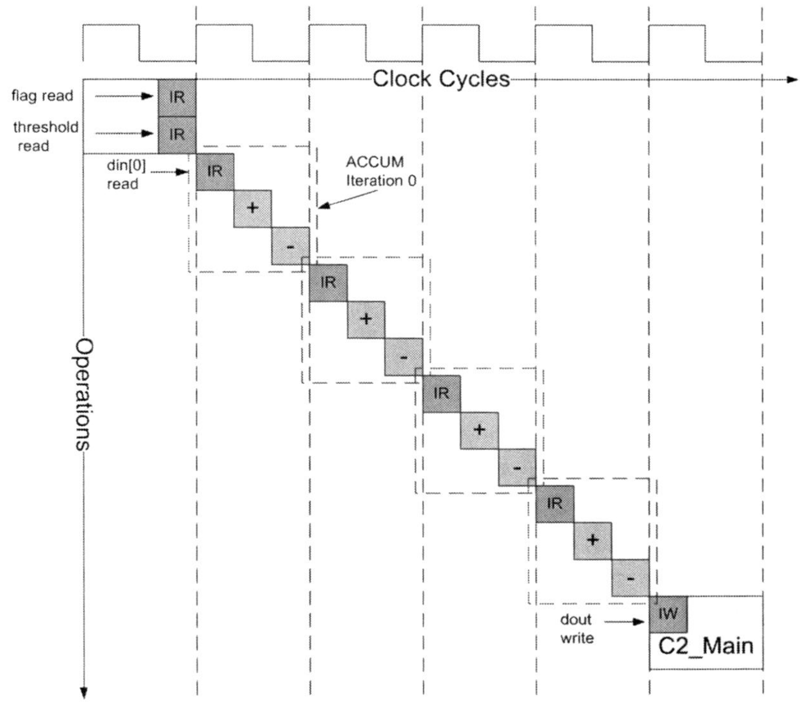

Figure 5-4 shows the resulting hardware implementation and off-chip configuration. Holding registers have been created for threshold and flag.

Because of this the hardware does not require that these variables are held stable through all iterations of the ACCUM loop. However, this design still has the same synchronization issues see when passing by reference. The off-chip hardware must guarantee that threshold and flag are setup and available for reading in C1 of the main loop without receiving any hardware synchronization signals from the on-chip logic. This is one of the reasons why "wire" interfaces are used either for designs that have IO read and written every clock cycle, or in combination with some hardware synchronization to make sure that IO is accessed in the correct c-step.

Figure 5-4. Hardware of Unconditional IO Passed by Value

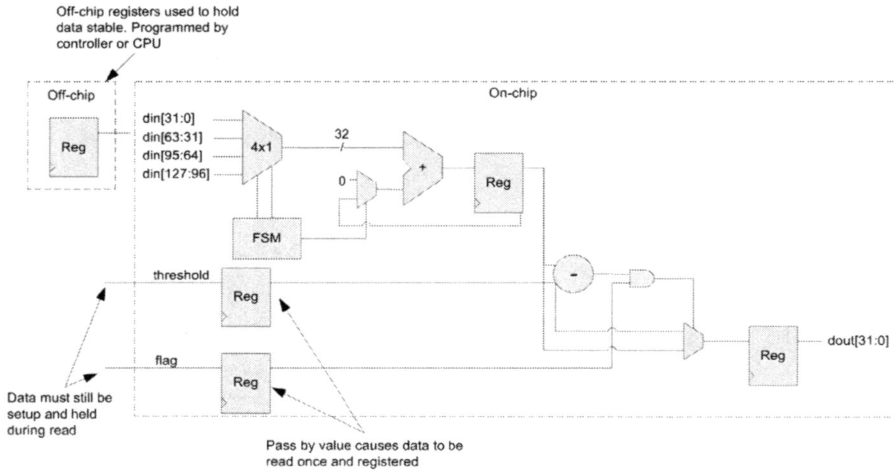

Conditional IO

An IO is considered conditional if the interface variable is mapped to a resource that has a hardware "handshake". This handshake can consist either of a simple ready to send or receive data, or a ready/acknowledge behavior. Unlike unconditional IO, where high-level synthesis is free to move IO into and out of conditions in the C++, conditional IO cannot be moved into or out of conditions in the C++ code. The only exception to this rule is when the variable mapped to IO is pass-by-value. In this case the IO is always read once at the beginning of the design schedule and stored in registers. Using pass-by-value variables in this way can have some potentially unexpected behavior when the IO has a handshake.

Pass by Reference

Similar to the pass-by-reference example for unconditional IO, passing by reference using conditional IO requires that the data is setup and available before the IO is accessed. However conditional IO provides the mechanism to synchronize the transfer of data via ready/acknowledge control signals. In Example 5-3 the “threshold” variable is mapped to an IO resource that generates a ready for data strobe. It is assumed that the data is already available for reading in an off-chip FIFO so an acknowledge is not required. The “threshold” interface variable is read conditionally after the “flag” interface variable is read and evaluates to true.

Example 5-3. Conditional IO Passed by Reference

```
void accumulate(int din[4], int &dout, int &threshold, bool &flag)
{
  int acc=0;
 ACCUM:for(int i=0;i<4;i++){
    acc += din[i];
    if(flag)
      if(acc > threshold)
        acc = threshold;
  }
  dout = acc;
}

Design constraints
ACCUM loop with II=1
threshold and dout mapped to ready to send or receive data
interface
All other IO mapped to wire interfaces
```

Figure 5-5 shows the schedule for Example 5-3. The IO for “threshold is considered synchronous because the “ready for data” control signal is a registered signal. This is indicated in the schedule by showing the IO operation crossing the clock boundary.

Figure 5-5. Schedule of Conditional IO Passed by Reference

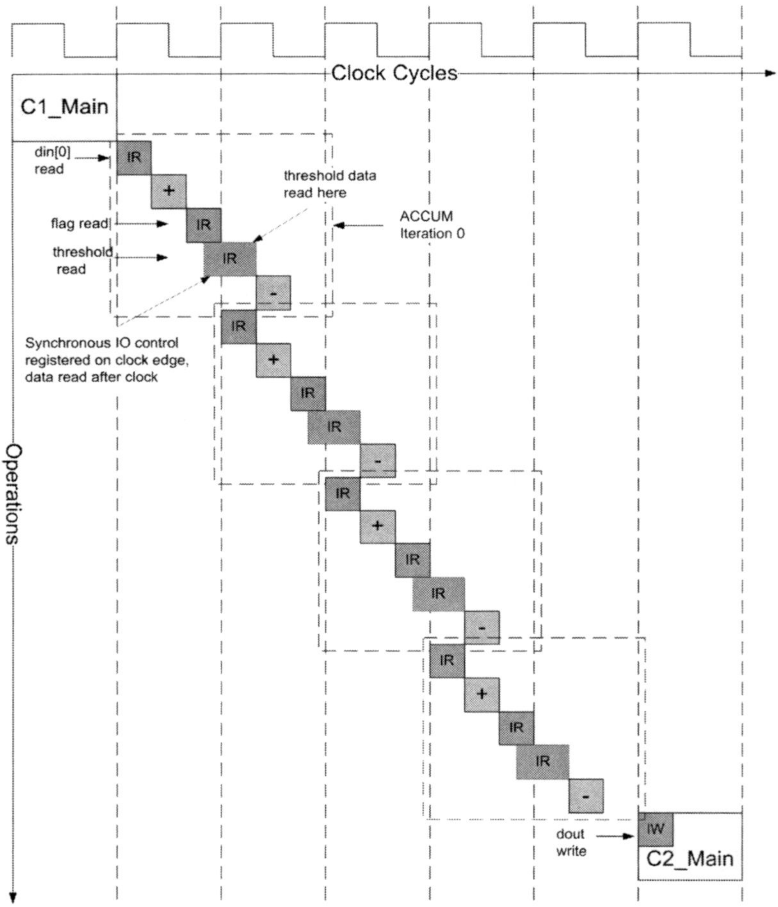

One of the effects of reading "threshold" inside of a condition based on "flag" is that the number of c-steps for each iteration has been increased. The "threshold" request-for-data control signal cannot be asserted until the condition based on "flag" has evaluated to true. The timing diagram of this behavior is shown in Figure 5-6. The read request for threshold is only generated when "flag" is asserted high.

Figure 5-6. Timing of Conditional IO Passed by Reference

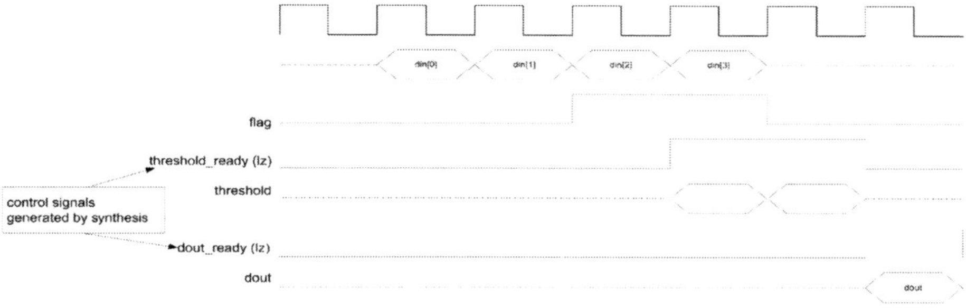

Figure 5-7 shows the hardware implementation of Example 5-3. Mapping to an IO with a "request" signal causes the synthesis process to insert a hardware control signal that can be hooked up to an external FIFO to control the flow of data. Additionally, the synchronous nature of "threshold" requires that "flag" is read in the previous clock cycle than "threshold".

Figure 5-7. Hardware of Conditional IO Passed by Reference

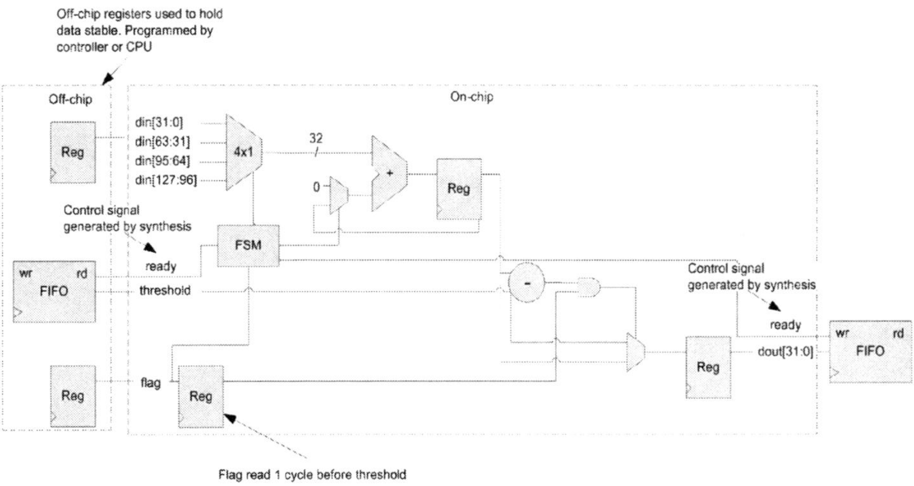

Pass by Value

The previous section illustrated how a pass-by-reference interface variable mapped to a conditional IO is only read inside of a C++ condition when the condition is true. This includes the generation of the "request" control signal. When using pass-by-value variables on the interface the behavior is different. Pass-by-value interface variables are always read at the beginning of the main loop regardless of where they are used in the C++ code. So even if the variable is read conditionally in the code, as shown in Example 5-4, a request-for-data is still generated.

Example 5-4. Conditional IO Passed by Value

```
void accumulate(int din[4], int &dout, int threshold, bool flag)
{
  int acc=0;
 ACCUM:for(int i=0;i<4;i++){
    acc += din[i];
    if(flag)
      if(acc > threshold)
        acc = threshold;
  }
  dout = acc;
}

Design constraints
ACCUM loop with II=1
threshold and dout mapped to request for data interface
All other IO mapped to wire interfaces
```

The schedule for Example 5-4 is shown in Figure 5-8. "threshold" has its request-for-data control signal issued in C1_Main and "threshold" and "flag" are read once, and only once, in C2_Main since they are both pass by value. The timing diagram in Figure 5-9 shows the potentially unexpected behavior of this example. The "threshold" request-for-data control signal is asserted and "threshold" is read is regardless of the value of "flag". This happens because pass-by-value interface variables are always read once at the beginning of the main loop no matter where they are used in the C++ code. Figure 5-10 shows the hardware synthesized for Example 5-4.

Figure 5-8. Schedule of Conditional IO Passed by Value

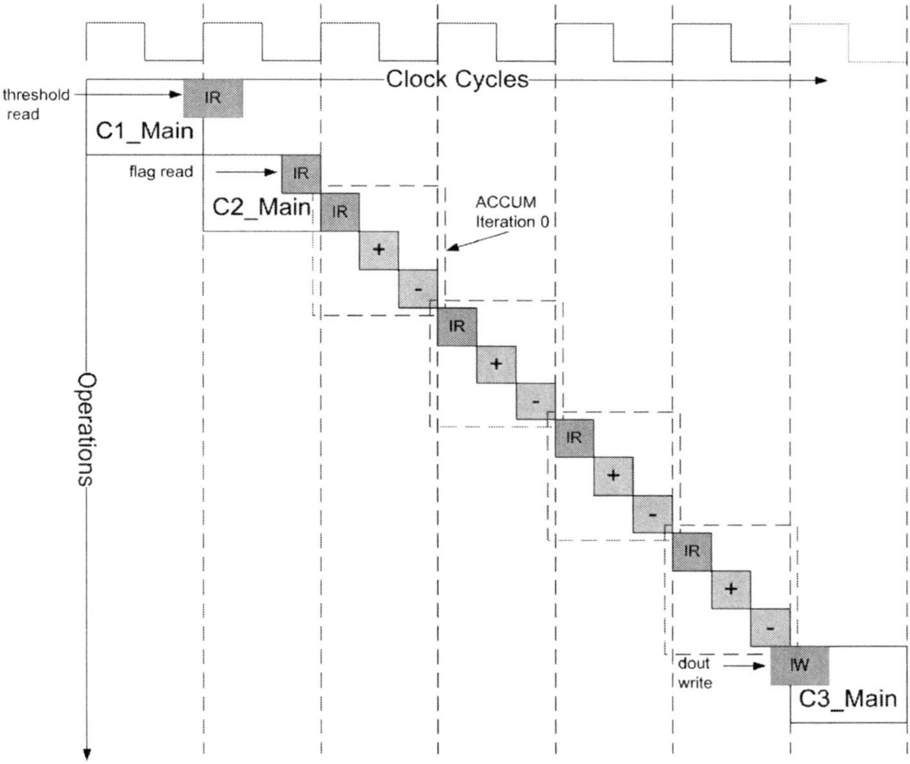

Figure 5-9. Timing of Conditional IO Passed by Value

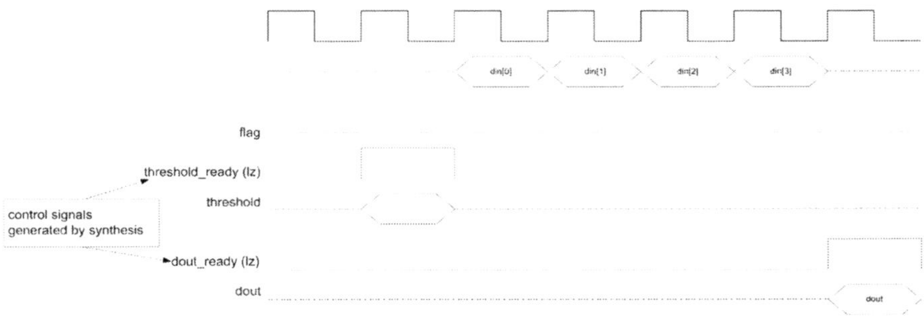

Figure 5-10. Hardware of Conditional IO Passed by Value

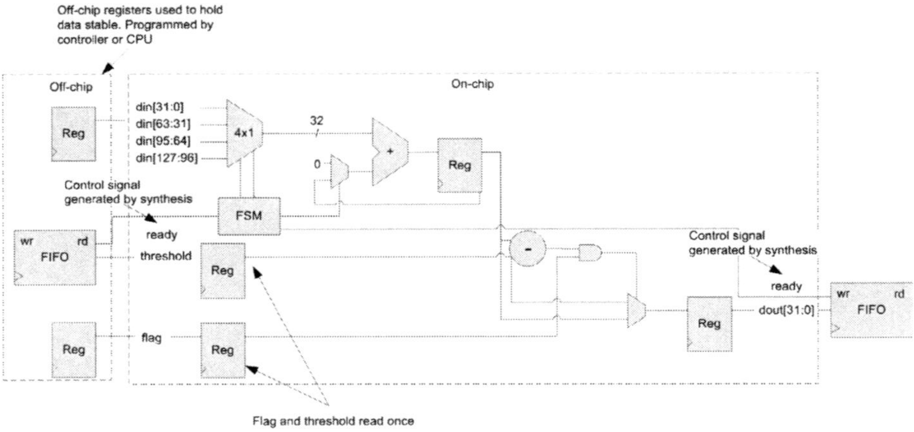

Ready/acknowledge Behavior (wait)

In addition to being able to automatically map interface variables to request-for-data type resources, high-level synthesis lets users map to interface resources that have a ready/acknowledge type behavior. These type of interfaces must be used with caution since they are more restrictive in terms of how the generated hardware behaves. In particular they can produce unwanted behavior when pipelining the main loop with II=1. This more restrictive case is discussed in the next section. For now we can re-use Example 5-3 on page 91 with a different set of constraints.

```
Design constraints
ACCUM loop with II=1
threshold mapped to request-grant interface
dout mapped to request for data interface
All other IO mapped to wire interfaces
```

Mapping the "threshold" interface variable to a ready/acknowledge type resource yields essentially the same schedule as that shown in Figure 5-5 on page 91. However the timing and hardware implementation is slightly different. The timing diagram for this example is shown in Figure 5-11. The "threshold" acknowledge control signal is used to determine when data is available. If data is being requested, "threshold" request driven high, and the acknowledge signal is low, the current loop iteration stalls the entire design until acknowledge goes high. This requires that the current data that is driven from "off-chip" must be held stable until the iteration completes. The timing diagram shows that this type of ready/acknowledge interface behavior is well suited for connecting to an off-chip FIFO, where ready is connected to the FIFO read, and acknowledge is connected to the FIFO empty flag, Figure 5-12.

Figure 5-11. Timing of IO with Wait

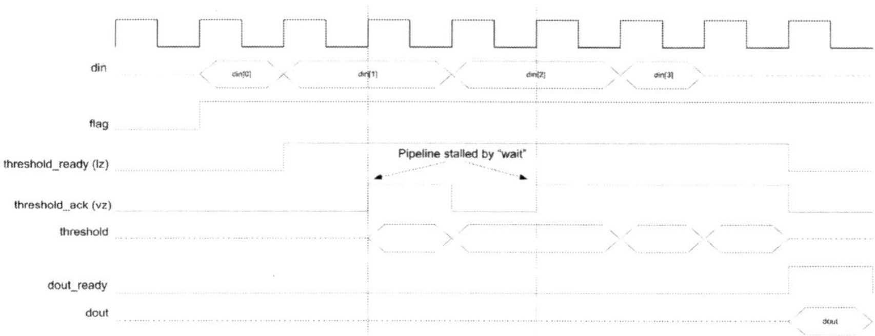

Figure 5-12. Hardware of IO with Wait

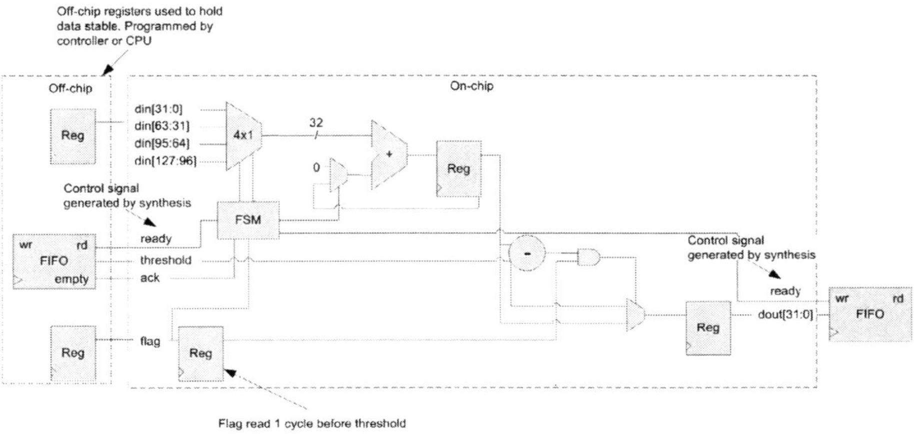

Stalling the Pipeline

Using IO resources with ready/acknowledge behavior showed that it is possible to stall the execution of a loop until data is available without any unwanted behavior. In the previous examples the main loop was left unconstrained while the ACCUM loop was pipelined with II=1. This allows the ACCUM loop to ramp-up and ramp-down, which in turn allows any data in the pipeline to "flush" after the last input is read. In designs with pipelines consisting of more than one stage, this "flushing" does not occur if the main loop is pipelined with II=1, and the whole pipeline can stall before the last output is written. Example 5-5 shown below is the four element accumulator that was used in previous sections.

Example 5-5. Stalling the Pipeline with Conditional IO

```
void accumulate4(int din[4], int &dout){
  int acc=0;
 ACCUM:for(int i=0;i<4;i++){
    acc += din[i];
  }
  dout = acc;
}

Design constraints
Main loop pipelined with II=1
ACCUM loop fully unrolled
din maped to ready-acknowledge resource
dout mapped to ready resource
```

Fully unrolling the ACCUM loop creates a two-stage balanced adder tree. In this example it is assumed that the clock frequency is sufficiently fast so that the adder tree stages are scheduled in separate c-steps. Figure 5-13 shows the schedule for Example 5-5. This shows that the read for iteration 1 of the main loops happens before the write of iteration 0. If the read data is not available the pipeline stalls.

Figure 5-13. Schedule of Pipelined Main Loop with Conditional Wait IO

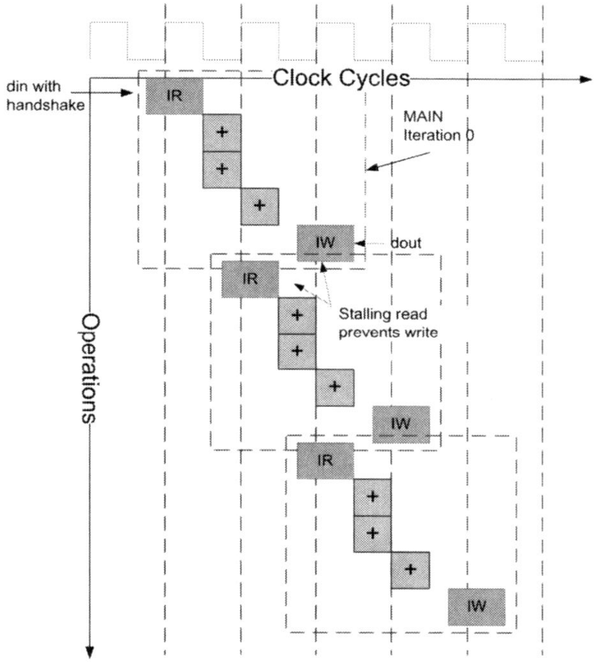

Figures 5-14 and 5-15 show the timing and hardware diagrams for Example 5-5. They illustrate how the previous read is completing in pipeline stage 2 while the current read is needed for pipeline stage 1. If the current read data is unavailable the previous read data gets stuck in the pipeline. In other words, the pipeline does not flush if there is no data available for reading at the input.

Figure 5-14. Timing of Pipeline Stall

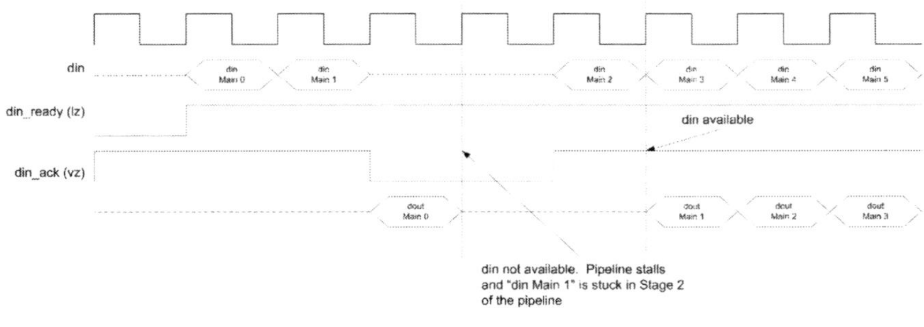

Figure 5-15. Hardware of Pipelined Main Loop with Conditional Wait IO

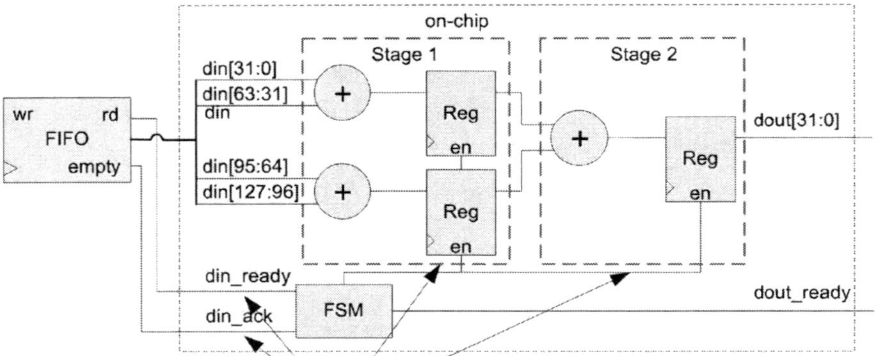

Having the pipeline stall is sometimes unacceptable for certain types of designs, especially designs that do not have continuously running data. Video is a good example of this, where horizontal and vertical blanking create gaps in the pixel data. One way to prevent the pipeline from stalling is to not pipeline the main loop but pipeline the inner loops. Pipelining the inner loops allows the pipeline to ramp down, flushing all data. The downside of not pipelining the main loops is that there is a multi-clock cycle penalty for the time it takes to ramp-up and ramp-down the pipeline. If the main loop must be pipelined for performance reasons the other solution is to manually code the "ack" into the C++ code to allow the pipeline to flush.

Caution

Pipelining the main loop when using handshaking IO can prevent the pipeline from flushing.

Manually Flushing the Pipeline

It's possible to manually flush the pipeline in a design with a pipelined main loop by explicitly coding the acknowledge into the C++ interface. Example 5-6 has taken the code from Example 5-5 and modified it so that acknowledge is now part of the top-level C++ interface.

Example 5-6. Manually Flushing the Pipeline

```
void accumulate(int din[4], int &dout, bool &ack){
  int acc=0;
  if(ack){
  ACCUM:for(int i=0;i<4;i++){
      acc += din[i];
    }
    dout = acc;
  }
}

Design constraints
Main loop pipelined with II=1
ACCUM loop fully unrolled
din maped to ready for data resource
dout mapped to ready to send resource
```

The C++ is written so that an output is always produced every time "ack" is true. When "ack" is false the loop is skipped and the function exits. This behavior allows the pipeline to flush since the design doesn't have to wait if data is not available. One side effect of this is that the read-for-data signal for "din" is not asserted until "ack" goes high. In the previous example that mapped "din" to a ready/acknowledge interface the ready-for-data signal was issued immediately regardless of the value of the acknowledge. Figure 5-16 shows the timing of Example 5-6.

Figure 5-16. Timing of Manually Flushing the Pipeline

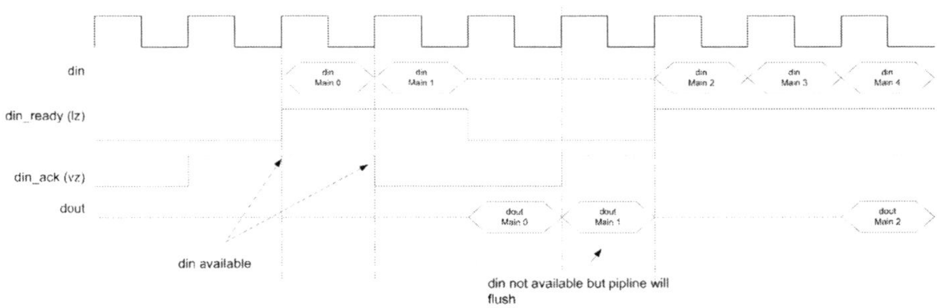

Writing IO for Throughput

All of the IO examples covered previously have been able to schedule with the main loop pipelined with II=1. This was because either there was only a single IO access per loop iteration or because the IO accesses were automatically merged when inside of an unrolled loop. There are instances when multiple IO accesses or conditional IO inside of unrolled loops are not merged, which in turn prevents pipelining a design and limits the throughput. When this happens it usually requires modifying the C++. Example 5-7 shows the four element accumulator with a slight modification. The “flag” array is tested inside the loop to determine which element of “din” is read and accumulated. If all elements of “flag” are false then “din” is never read. This design cannot be pipelined when the ACCUM loop is unrolled because multiple IO access are created and cannot be merged.

Example 5-7. IO Throughput Limiting Design

```
void accumulate(int din[4], int &dout, bool flag[4]){
  int acc=0;
  ACCUM:for(int i=0;i<4;i++){
      if(flag[i])
        acc += din[i];
  }
  dout = acc;
}

Design constraints
ACCUM loop fully unrolled
One ADD takes most of
din maped to ready for data resource
dout mapped to ready to send data resource
```

Figure 5-17 shows the schedule for Example 5-7where the main loop is not pipelined. Reading the IO “din” conditionally, where each condition is different “flag[i], in this case causes four separate conditional reads that are not merged, even though each read is only accessing a slice of “din”. This can be better understood by looking at Example 5-8 which shows

Figure 5-17. Schedule of IO Throughput Limiting Design

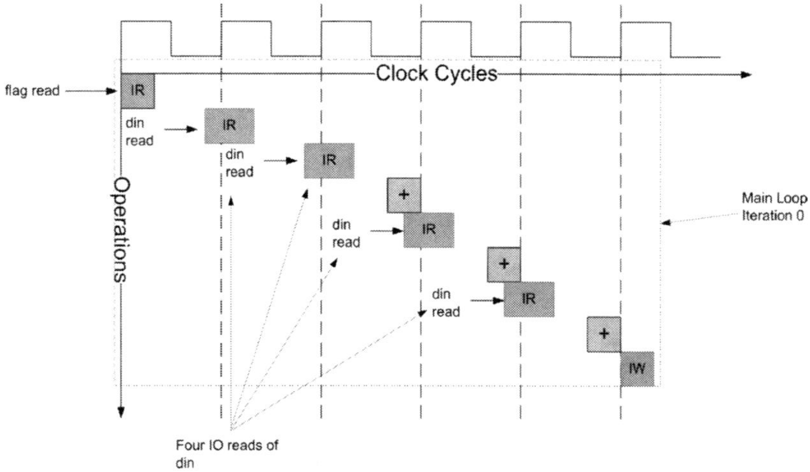

Example 5-7 with the ACCUM loop manually unrolled. What this illustrates is that each condition must be evaluated before "din" can be read and accumulated. Thus there is a dependency between each loop iteration and this prevents the IO accesses from being merged into a single access, causing four separate reads of "din".

Example 5-8. Manual Unrolling of IO Throughput Limiting Design

```
void accumulate(int din[4], int &dout, bool flag[4]){
  int acc=0;

  if(flag[0])
    acc += din[0];
  if(flag[1])
    acc += din[1];
  if(flag[2])
    acc += din[2];
  if(flag[3])
    acc += din[3];

  dout = acc;
}
```

The reason why this design cannot be pipelined is evident from Figure 5-18 which shows that overlapping iterations of the main loop would require simultaneous multiple reads from the IO port "din", which is physically impossible.

Figure 5-18. Design that Can't be Pipelined Due to Unmerged IO

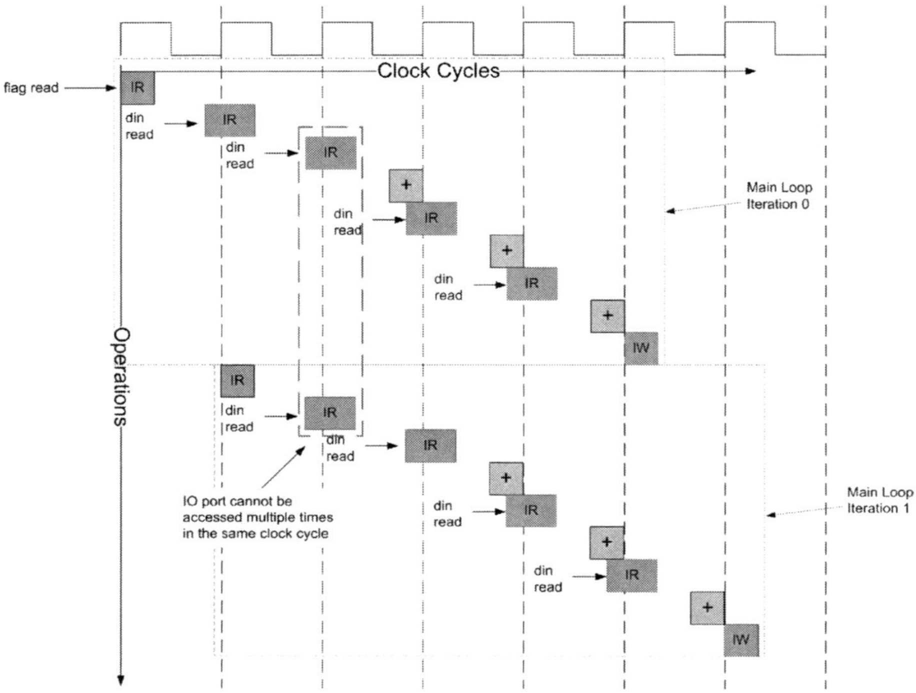

Making IO Mergable

The code for Example 5-7 should be rewritten to either make the read of "din" unconditional when possible, or to simplify the condition so that the reads can be merged. Example 5-7 illustrates how you often get what you asked for, but not what you want, when writing synthesizable C++ code. Let's assume that after analyzing the undesirable scheduling results from HLS, it is determined that "din" can, and should, be read every iteration of the main loop, since in hardware it is expected that all four values of "din" come in parallel from an external FIFO. With this assumption the C++ code can be rewritten as shown in Example 5-9. All elements of "din" are read in the beginning of the design regardless of the value of "flag", and then stored in the internal variable "din_int". The internal variable is then used in the ACCUM loop.

Example 5-9. Making IO Read Unconditional

```
void accumulate(int din[4], int &dout, bool flag[4]){
  int acc=0;
  int din_int[4];
  bool flag_int;

 DIN:for(int i=0;i<4;i++)
    din_int[i] = din[i];
 ACCUM:for(int i=0;i<4;i++){
    if(flag[i])
      acc += din_int[i];
  }
  dout = acc;
}

Design constraints
Main lop pipelined with II=1
DIN and ACCUM loops fully unrolled
din maped to ready for data resource
dout mapped to ready to send resource
```

Figure 5-19 shows the scheduled design for Example 5-9 where making the read of all elements of "din" unconditional allows them to be merged into a single read, which in turn allows the design to be pipelined with II=1. Although the reads have been merged, the three adders in the design have a dependency because of the conditional accumulate based on "flag[i]". This dependency prevents adder tree balancing and can result in sub-optimal hardware, especially as the clock frequency is increased to the point where the adders must be scheduled in separate clock cycles.

Figure 5-19. Schedule of Unconditional IO Read

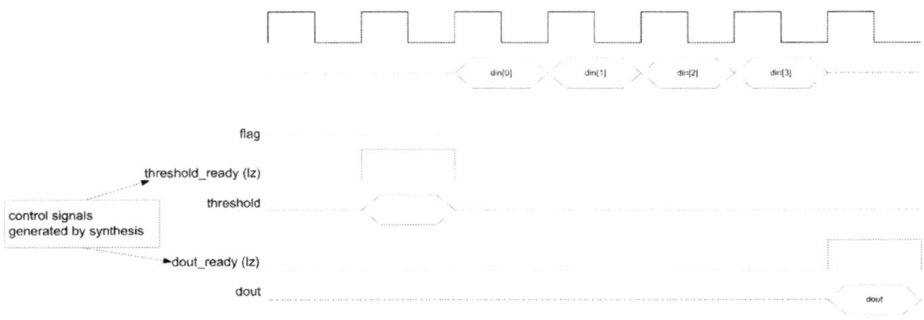

One potential problem with the re-write of Example 5-9 is that the read of "din" always occurs regardless of whether any of the "flag[i]" elements are set, which is different from the behavior of Example 5-7.

If the desired behavior is to only read "din" if at least one element of "flag[i]" is set the code can be rewritten to give this type of behavior, shown in Example 5-10.

Example 5-10. Simplifying Conditional IO to Help Merging

```
void accumulate(int din[4], int &dout, bool flag[4]){
  int acc=0;
  int din_int[4];
  bool flag_int;
 FLAG:for(int i=0;i<4;i++)
    flag_int |= flag[i];
 DIN:for(int i=0;i<4;i++)
    if(flag_int)
      din_int[i] = din[i];
 ACCUM:for(int i=0;i<4;i++){
    if(flag[i])
      acc += din_int[i];
    else
      acc += 0;
  }
  dout = acc;
}

Design constraints
Main loop pipelined with II=1
FLAG, DIN and ACCUM loops fully unrolled
din maped to ready for data resource
dout mapped to ready to send resource
```

Example 5-10 has made the read of "din" conditional by creating a boolean variable that is equal to the "OR" of all of the "flag[i]" elements, which is done in the FLAG loop. If "flag_int" is true then "din" is read. Using the simple condition inside of the DIN loop allows the IO reads to be merged.

Caution

Conditionally reading arrays mapped to IO inside of unrolled loops has the potential to prevent pipelining. Make the IO reads unconditional when possible by reading the entire array into an internal array.

Memories

HLS not only allows users to map arrays to IO resources, where the array elements are available in parallel with or without a handshake, but also allows them to map arrays to memory type resources. Both internal arrays and arrays on the top-level function interface can be mapped to memory resources. If the array is on the top-level interface HLS creates the address, data, and control signals required to interface to an off-chip memory. If the array is internal to the design HLS not only creates the necessary address, data, and control signals to access the memory, but it also instantiates the memory model. In the case of targeting ASIC designs this instantiation of the memory is only used for simulation, and is black boxed for synthesis since ASIC synthesis does not infer memories. The "black-box" memory can then be replaced with the physical memory produced by the users memory compiler. In the case of FPGA design, the instantiated memory is used not only for simulation, but is inferred as a memory by the FPGA RTL synthesis tool.

Automatic Mapping of Arrays to Memories

Consider the following four element accumulator example used previously where the interface array is now mapped to a memory resource.

Example 5-11. Arrays Mapped to Memories

```
void accumulate(int din[4], int &dout){
  int acc=0;
 ACCUM:for(int i=0;i<4;i++){
    acc += din[i];
  }
  dout = acc;
}

Design constraints
Main loop pipelined with II=1
din maped to single port RAM interface
dout mapped to ready to send resource
```

Leaving the ACCUM loop left rolled in Example 5-11, with the main looped pipelined with II=1, requires four clock cycles, or four 32-bit memory reads, to read the data from the memory and write the output to "dout", as shown in Figure 5-20. One important thing to note here is that there is only one memory read per iteration of ACCUM, which allows the design to be pipelined with II=1. Figure shows the timing diagram. HLS always uses synchronous memories so the address is issued in the clock cycle prior to the data when reading. For memory writes address and data are issued in the same cycle.

Figure 5-20. Arrays Mapped to Memories

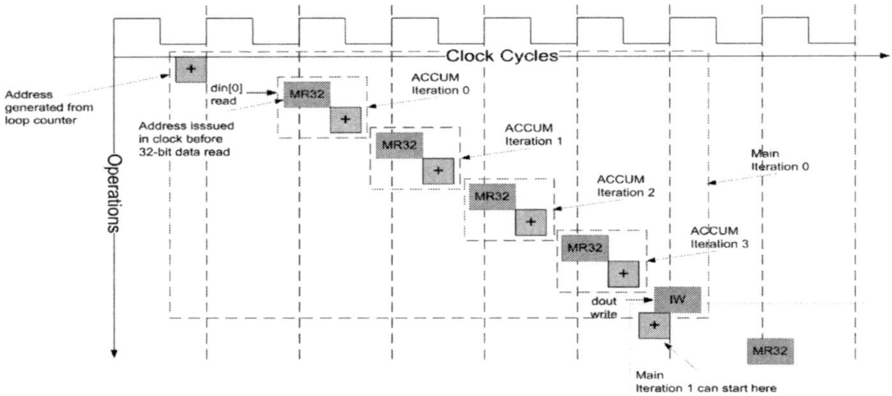

Figure 5-21. Timing of Arrays Mapped to Memories

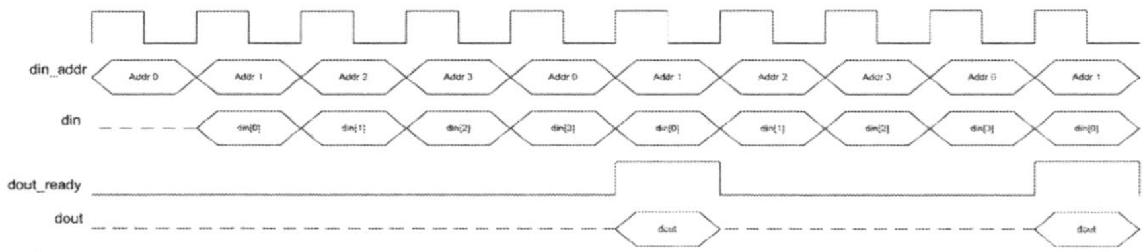

Figure shows the hardware diagram for Example 5-11. The "din" array, which is mapped to a memory-type interface, is synthesized to hardware that contains an address/data memory interface that allows the design to be hooked up to an off-chip memory.

Figure 5-22. Hardware of Arrays Mapped to Memories

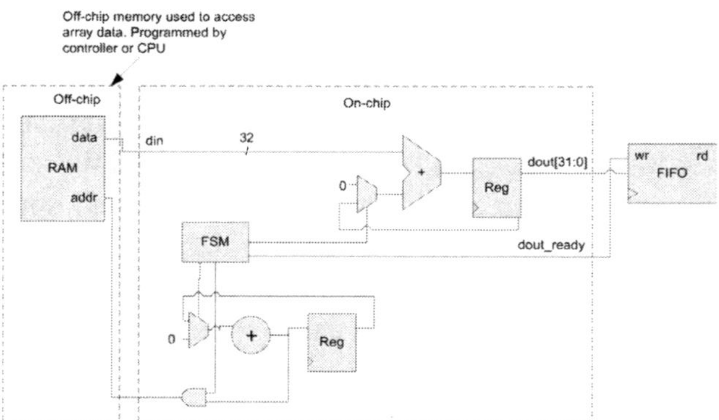

Automatic Memory Merging

Similar to IO, the way in which arrays mapped to memories are accessed in the C++ code affects performance, as well as HLS's ability to automatically optimize memory accesses. One of the more powerful features of HLS is automatic memory merging, where sequential reads and writes to memories can be combined when the width of a memory is doubled, tripled, etc. If memory accesses cannot be automatically merged by HLS the C++ code must be re-written either to facilitate merging, or to manually merge the memory accesses.

Some of the conditions for automatic memory merging to happen are:

- Reads and writes to arrays mapped to memories must start on even word boundaries.
- Reads and writes to arrays mapped to memories should be unconditional when possible.
- Conditional reads and writes to memories inside of unrolled loops should use simple conditions to allow merging.
- Multiple reads and writes to arrays mapped to memories within the same loop body must be in mutually exclusive conditions.
- The number of elements of one dimensional arrays mapped to memories must be divisible by the factor in which the memory is automatically widened if memory merging is to occur.
- When using two dimensional arrays the right-most dimension must be a power of two.

The hardware architecture shown previously in Figure 5-22 is limited to reading one 32-bit value from the memory interface per clock cycle. The physical memory interface is usually the bottleneck in the performance of the algorithm because the internal bandwidth of the design cannot exceed the interface bandwidth. In Example 5-11, the internal and interface bandwidths were exactly matched, where each iteration of the ACCUM loop read one 32-bit value from the memory mapped array “din” on the design interface. What we saw previously is that loop unrolling can be used as a mechanism to increase design performance for arrays mapped to wire type interfaces. This was because, in most cases, anywhere from one to all elements of the array could be read at once from the interface. This is generally not possible when arrays are mapped to memories. In order to understand why, it's helpful to look at the effects of loop unrolling on the schedule when arrays are mapped to memories. Consider Example 5-11 with the following constraints.

```
Design constraints
din maped to single port RAM interface
ACCUM loop unrolled by 2
dout mapped to ready to send resource
```

Figure 5-23 shows the schedule and the effects of unrolling the ACCUM loop. Although the number of loop iterations has been halved by unrolling by two, the performance has not improved since it still requires the same number of clock cycles to read from the single port memory. In fact the performance of this version of the design is worse than the rolled loop version because this version cannot be pipelined with II=1. Figure 5-24 illustrates why this is not possible.

Figure 5-23. Unmerged Memory Accesses Inside Unrolled Loops

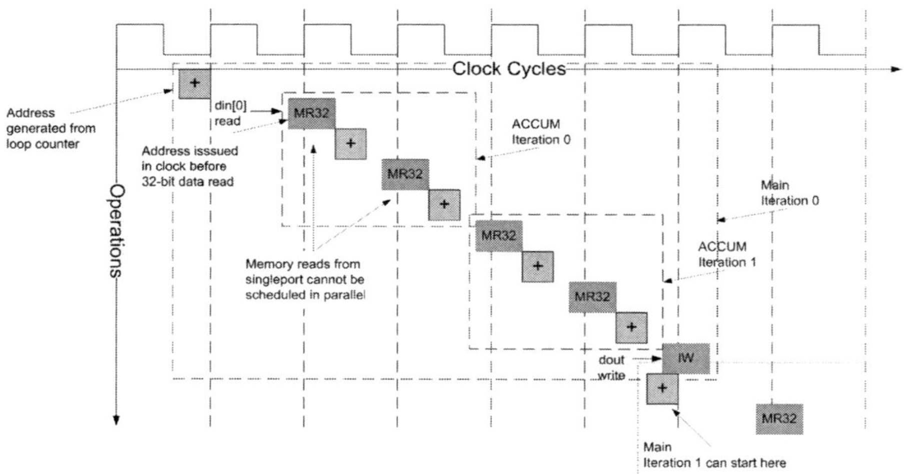

Trying to pipeline the design with II=1 essentially means that a new iteration of the ACCUM loop, which has been flattened into the main loop (See "Nested Loops" on page 56), must be started every clock cycle. However this is not possible because it would require reading twice from the singleport memory in the same clock cycle. In other words this design cannot be pipelined with II=1 for the schedule shown in Figure 5-23.

Figure 5-24. Failed Schedule for Unmerged Memory Accesses with II=1

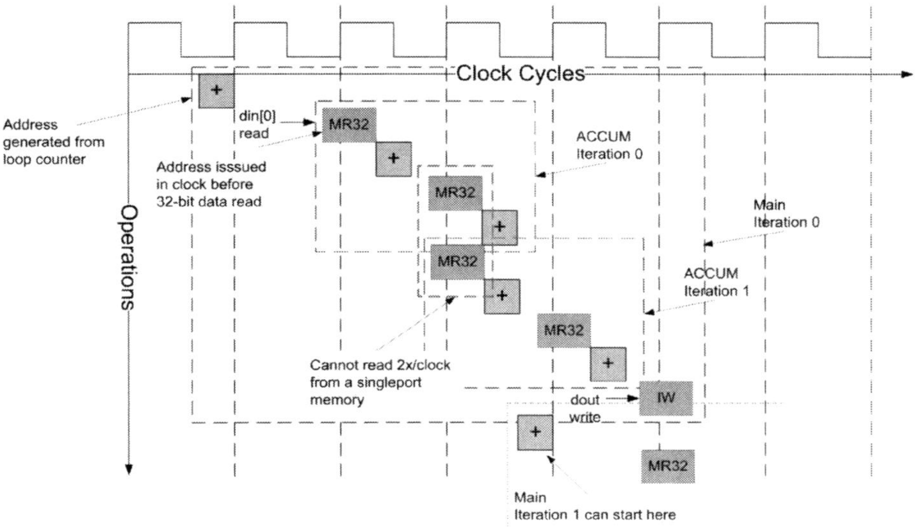

The memory reads to the singleport memory must be reduced to one read per loop iteration in order to schedule the design when pipelined with II=1. For this example this can be achieved by automatically widening the word width of the memory interface to 64 bits. Since the reads of "din" begin on an even word boundary, automatic memory merging should be able to combine them into a single read per loop iteration. The design is re-constrained with the following set of constraints:

```
Design constraints
Main loop pipelined with II=1
din maped to single port RAM interface
Word width of singleport interface widened to 64 bits
ACCUM loop unrolled by 2
dout mapped to ready to send resource
```

Figure 5-25 shows the scheduled design after the word width of the memory interface is doubled, allowing the two 32-bit sequential reads to "din" to be merged into a single 64-bit read.

Figure 5-25. Schedule for Merged Memory Accesses with II=1

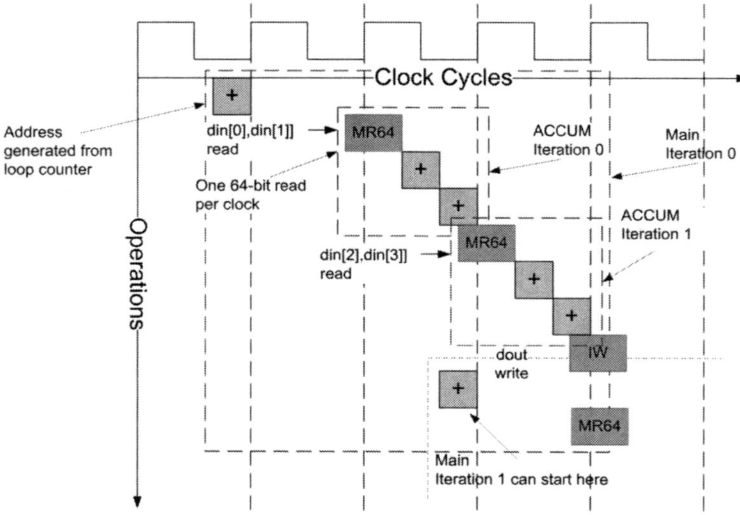

Designing for Throughput When Using Memories

The previous examples with arrays mapped to singleport memory interfaces have illustrated that only one read or write to a singleport memory per clock cycle is possible. This is an important aspect of HLS to pay attention to when structuring the C++. If an array is read or written at different places within the design it is essential that the code expresses mutual exclusivity of the array/memory accesses, or the code should be rewritten to have only a single array/memory access.

Non-Mutually Exclusive Memory Accesses

Consider the following design in Example 5-12. Two interface variables "flag0" and "flag1" are used to control whether the elements of "din" are added or scaled and subtracted. Even if "flag0" and "flag1" are never set at the same time, it is impossible for HLS to prove this, and two memory reads are scheduled. This design cannot be pipelined with II=1 due to the multiple reads on "din". Figure 5-26 shows the failed schedule of the design to illustrate why pipelining is not possible.

Example 5-12. Non-Mutually Exclusive Memory Accesses

```
void accumulate(int din[4], int &dout, bool &flag0, bool &flag1)
{
  int acc=0;
 ACCUM:for(int i=0;i<4;i++){
    if(flag0)
      acc += din[i];
    if(flag1)
      acc -= din[i]/2;
  }
  dout = acc;
}

Design constraints
Main loop pipelined with II=1
din maped to single port RAM interface
dout mapped to ready to send resource
```

Figure 5-26. Failed Schedule for Non-mutually Exclusive Memory Accesses

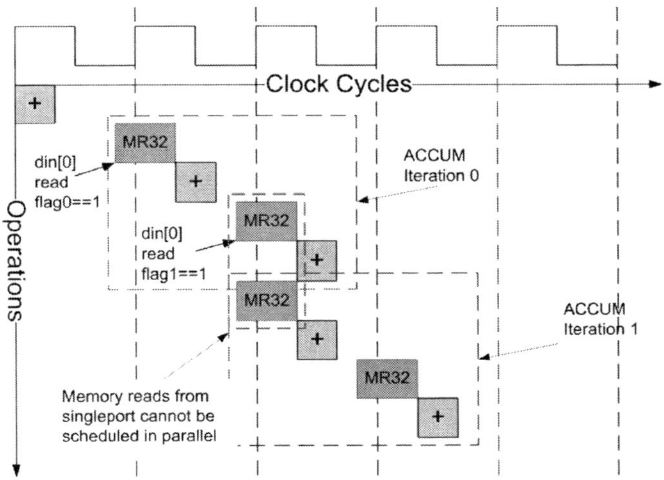

Figure 5-26 shows that pipelining with II=1 is not possible since each iteration of the ACCUM loop requires two reads from the singleport memory interface. The C++ code must be re-written in order to pipeline this design.

Making Memory Accesses Mutually Exclusive

When possible, multiple accesses to an array mapped to memory should be made mutually exclusive. In Example 5-12 the C++ was written such that it is impossible for HLS to prove mutual exclusivity. The reads of "din" can be made to be mutually exclusive if it is known by the designer that "flag0" and "flag1" can never be true at the same time. Example 5-13 shows the re-written code with explicit mutual exclusivity by using an "if-else" statement instead of two "if" statements. Accessing "din" in separate branches of a condition allows the two reads to be merged into a single read operation. Figure 5-27 shows the schedule for Example 5-13. Because the address/index for "din[i]" is the same in both branches, it is merged into a single adder.

Example 5-13. Mutually Exclusive Memory Accesses

```
void accumulate(int din[4], int &dout, bool &flag0, bool &flag1)
{
  int acc=0;
 ACCUM:for(int i=0;i<4;i++){
    if(flag0)
      acc += din[i];
    else if(flag1)
      acc -= din[i]/2;
  }
  dout = acc;
}
```

Figure 5-27. Schedule of Mutually Exclusive Memory Accesses

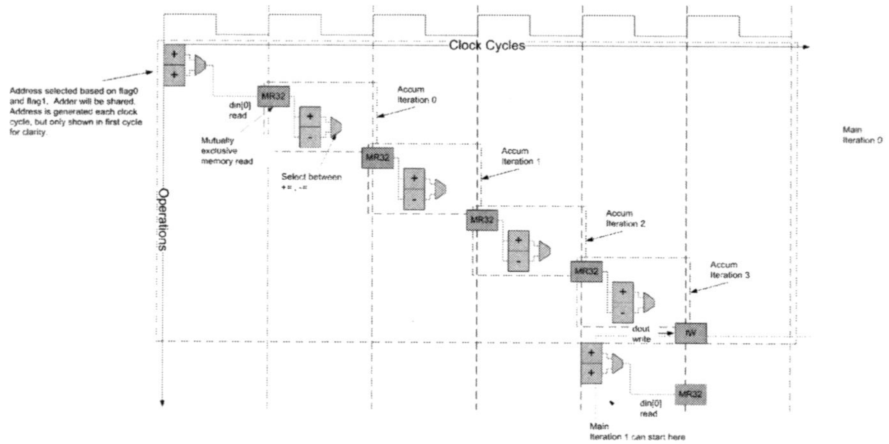

Manually Merging Non-Mutually Exclusive Memory Accesses

Example 5-13 showed how Example 5-12 can be rewritten to make memory accesses mutually exclusive when the conditional logic controlling the memory accesses is also known to be mutually exclusive. A different approach is required when the conditional accesses are not mutually exclusive. If the "flag" variables of Example 5-12 can both be true at the same time both conditions are entered. In this case the best approach is to try and manually reduce the number of reads of "din" to once per loop iteration. Example 5-14 shows the rewritten design where a temporary variable "tmp" has been used to read "din" once per loop iteration. "tmp" is then used internally. The design can now be pipelined with II=1. This code transformation was possible because the original design accessed the same address for both reads of "din". This technique would not help if the addresses were different and would require a different type of transformation. This type of transformation is covered in the chapter on memory architecture.

Example 5-14. Manually Merging Non-Mutually Exclusive Memory Accesses

```
void accumulate(int din[4], int &dout, bool &flag0, bool &flag1)
{
  int acc=0;
  int tmp;
 ACCUM:for(int i=0;i<4;i++){
    tmp = din[i];
    if(flag0)
      acc += tmp;
    if(flag1)
      acc -= tmp/2;
  }
  dout = acc;
}
```

Chapter 6

Sequential and Combinational Hardware

Introduction

The previous chapters provided a good introduction to the principles behind high-level synthesis and the use of bit-accurate data types. The basics of scheduling and loop optimizations were illustrated using concepts familiar to RTL designers, such as hardware diagrams, state machines, and timing diagrams. The next logical step is to take this foundation and apply it to some real world hardware examples. In a similar fashion to most RTL design guides, this chapter presents many of the basic hardware structures that RTL designers are familiar with, and shows how to code them using synthesizable C++. Unlike the examples of previous chapters, which focused primarily on a C-like coding style, class-based/object oriented C++ is introduced, including templates and recursion. As this chapter progresses the reader can begin to see the true power of hardware design using C++. The hardware examples presented in this chapter are all depicted as sequential circuits. This is because it is assumed that each example is synthesized as the top-level design. When used in the context of a larger design these circuits may be sequential or combinational based on the clock frequency and how the design is scheduled. True combinational components can also be synthesized by using explicit directives in the C++ synthesis tool.

Shift Registers

After going through the previous two chapters readers should have a good understanding of how HLS relates to some basic hardware structures. We can now build on that understanding by looking at some of the most commonly used structures in RTL design. All RTL designers are familiar with shift registers, and their many different variations, so this is a good place to begin describing basic hardware concepts using C++.

Basic Shift Register

Shift registers are used in a number of applications, with perhaps the most common use being in Finite Impulse Response (FIR) filters. The shift register consists of a chain of d-type flops that stores a history of values applied to the shift register input. Figure 6-1 shows the hardware diagram of a basic four-tap shift register. Every clock cycle a new value of "din" is read and stored, while the previously stored values of din are shifted to the right. In other words the shift register keeps a history of previous values of din, with the oldest value stored in the right-most register.

Example 6-1 shows a C++ function that implements a shift register. It uses a user-defined data type "dType" as well as a user-define constant "N_REGS" to define the number of shift register registers. A static array "regs" is declared internal to the function. The "static" declaration is required so that the past values of "din" stored in "regs" persist between calls to the shift_reg function. The SHIFT loop is used to do the actual shifting of data, and the WRITE loop is used to copy the shift register values to the output port "dout". One of the drawbacks of using this "C" like coding style for creating shift registers is that the function cannot be reused if multiple shift registers are needed. The "regs" array is shared between multiple calls to the "shift_reg" function because it has been declared static. If multiple shift registers are required the user must either create a separate function for each shift register, or use C++ template functions or classes to uniqueify the implementation. This is covered in a later section.

Figure 6-1. Basic Shift Register

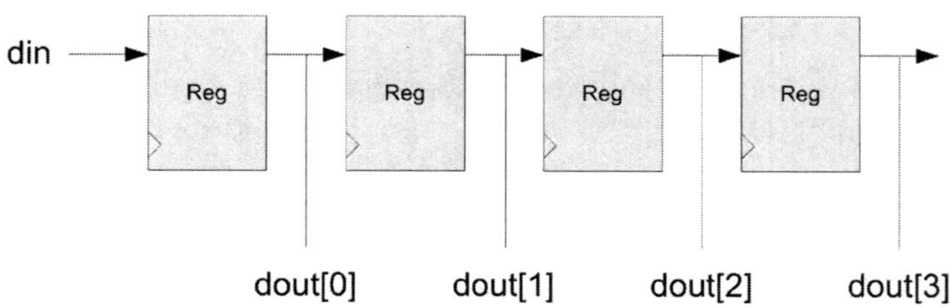

Example 6-1. Basic Shift Register

```
#include "basic_shift.h"
void shift_reg(dType din, dType dout[N_REGS]){
  static dType regs[N_REGS];
 SHIFT:for(int i=N_REGS-1;i>=0;i--){
    if(i==0)
      regs[i] = din;
    else
      regs[i] = regs[i-1];
  }
 WRITE:for(int i=0;i<N_REGS;i++)
    dout[i] = regs[i];
}

Design Constraints
Main loop pipelined with II=1
regs array mapped to registers
SHIFT and WRITE loops fully unrolled
```

Example 6-1 shows the relative ease in which C++ allows the description of hardware. One thing that should be noted with this example is that there is no clock, reset, enable, etc. Some of these signals such as the clock, clock enable, and resets can be added automatically during the synthesis process. Other signals such as data enable, load, and synchronous reset, can be defined in the C++ code.

Shift Register with Enable

The basic shift register from Example 6-1 can easily be enhanced to add a data enable signal to control the shifting of "regs", Example 6-2.

Example 6-2. Shift Register with Data Enable

```
#include "basic_shift.h"
void shift_reg(dType din, dType dout[N_REGS],bool en){
  static dType regs[N_REGS];
 SHIFT:for(int i=N_REGS-1;i>=0;i--){
    if(en){
      if(i==0)
        regs[i] = din;
      else
        regs[i] = regs[i-1];
    }
  }
 WRITE:for(int i=0;i<N_REGS;i++)
    dout[i] = regs[i];
}

Design Constraints
Main loop pipelined with II=1
```

```
regs array mapped to registers
SHIFT and WRITE loops fully unrolled
```

Example 6-2 shows that the boolean signal "en" is used to control whether or not the SHIFT loop body is executed. If "en" is true then the shift occurs, otherwise the values stored in "regs" are held. Figure 6-2 shows the hardware diagram for Example 6-2. The "en" signal that was added to the design causes a feedback MUX to be inserted at the input of each register to hold the output when "en" is false. This feedback MUX can then be transformed into a clock gate during the downstream RTL synthesis process.

Figure 6-2. Shift Register with Data Enable

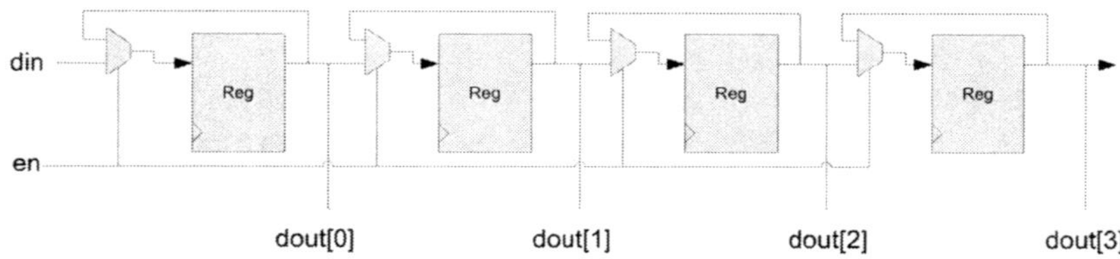

Shift Register with Synchronous Clear

Similar to adding a data enable signal, the shift register can be enhanced to allow clearing of all registers based on a control signal, shown below in Example 6-3. If the "srst" signal is true then all elements of "regs" are set equal to zero, otherwise the data is shifted. The reset in this case is synchronous because it is described within the C++, and HLS is always going to build a synchronous design.

Example 6-3. Shift Register with Synchronous Clear

```
#include "basic_shift.h"
void shift_reg(dType din, dType dout[N_REGS],bool srst){
  static dType regs[N_REGS];
 SHIFT:for(int i=N_REGS-1;i>=0;i--){
    if(srst)
      regs[i] = 0;
    else{
      if(i==0)
        regs[i] = din;
      else
        regs[i] = regs[i-1];
    }
  }
 WRITE:for(int i=0;i<N_REGS;i++)
```

```
      dout[i] = regs[i];
}

Design Constraints
Main loop pipelined with II=1
regs array mapped to registers
SHIFT and WRITE loops fully unrolled
```

Figure 6-3 shows the hardware diagram of Example 6-3.

Figure 6-3. Shift Register with Synchronous Reset

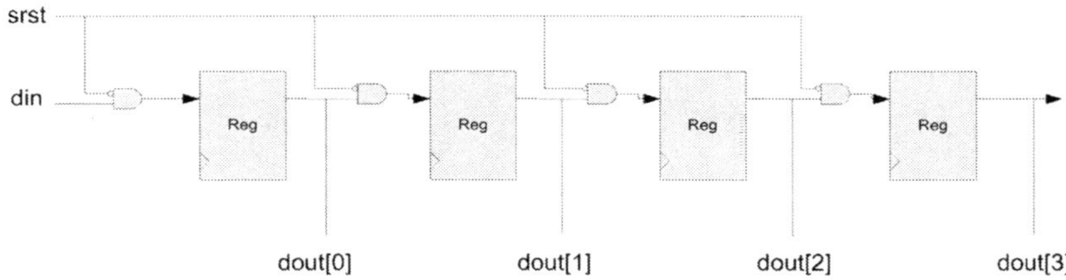

Shift Register with Load

A synchronous load can be added to the shift register to load "regs" from the design interface. This is done in the same fashion that the synchronous reset was added. Example show the shift register with a synchronous load. When "load" is true the elements of "regs" are loaded with the values of "load_data" read from the function interface. The hardware implementation for Example 6-4 is shown in Figure 6-4.

Example 6-4. Shift Register With Load

```
#include "shift_w_load.h"
void shift_reg(dType din, dType load_data[N_REGS],dType dout[N_
REGS],bool load){
  static dType regs[N_REGS];
 SHIFT:for(int i=N_REGS-1;i>=0;i--){
    if(load)
      regs[i] = load_data[i];
    else{
      if(i==0)
        regs[i] = din;
      else
        regs[i] =  regs[i-1];
    }
  }
 WRITE:for(int i=0;i<N_REGS;i++)
    dout[i] = regs[i];
}
```

```
Design Constraints
Main loop pipelined with II=1
regs array mapped to registers
SHIFT and WRITE loops fully unrolled
```

Figure 6-4. Shift Register with Load

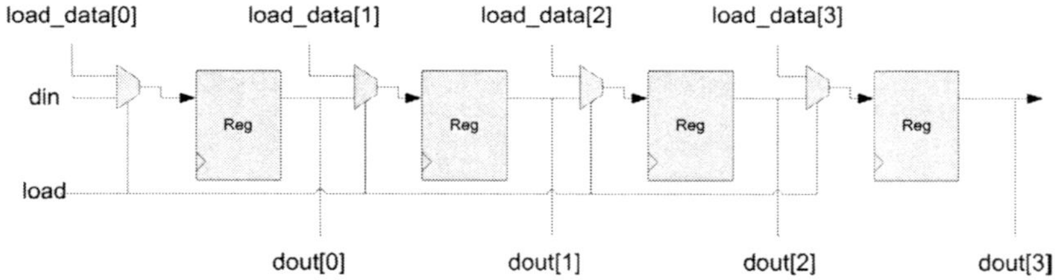

Shift Register Template Function

One of the problems with all of the shift register implementations covered so far is that the "shift_reg" function is not reusable. In other words it is not possible to create multiple instances of shift registers to be used within the same design. This is because the static declaration of the "regs" array is not unique, and is shared between all calls to the "shift_reg" function. This is essentially a limitation with the "C" language, but is easy to overcome using C++. Up to this point most of the design examples are more "C" like since they have not used many of the features of C++. C++ supports function templates that allow functions to not only operate with generic types, but can also allow function calls to be uniquefied. Templates are used in a design by using the C++ keyword "template" followed by one or more template arguments. Templates are similar to RTL generics or parameters, but they are much more powerful.

Example 6-5 shows a templatized version of the basic shift register. For this example there are three template parameters, "ID" which is an integer used to create a unique instance of the function, "dataType" which is used to specify the data type processed by the function, and NUM_REGS which controls the number of shift registers. It immediately becomes apparent that the function template has given us a design that can be reused. Lines 16 and 17 of Example 6-5 show how two unique instances of the shift_reg function can be created by specifying a unique value for the "ID" parameter. In addition to that, the data type and number of registers are also specified allow multiple unique instances of the shift register to be created for any data type and number of registers.

Example 6-5. Shift Register Function Template

```
#include "template_shift.h"
template<int ID, typename dataType, int NUM_REGS>
void shift_reg(dataType din, dataType dout[NUM_REGS]){
  static dataType regs[NUM_REGS];
 SHIFT:for(int i=NUM_REGS-1;i>=0;i--){
    if(i==0)
      regs[i] = din;
    else
      regs[i] = regs[i-1];
  }
 WRITE:for(int i=0;i<NUM_REGS;i++)
    dout[i] = regs[i];
}

void shift_reg_instances(int din0, char din1, int dout0[N_REGS0],
char dout1[N_REGS1]){
  shift_reg<1,int,N_REGS0>(din0,dout0);
  shift_reg<2,char,N_REGS1>(din1,dout1);
}

Design Constraints
Main loop pipelined with II=1
regs array mapped to registers
SHIFT and WRITE loops fully unrolled
```

Class Based Shift Register

Up to this point we've seen how a number of different shift register implementations can be realized using C++. C++ templates were introduced and illustrated the benefits of being able to reuse a design description for different data types and different number of registers. However, having to create separate implementations for all the different types of shift registers is undesirable. What is needed is a single description that can be configured to do what we want. While it is possible to use function templates to do this, a much better approach is to create a shift register class. This class not only is templatized for the data type and number of registers, but allows unique instances without the need for a template "ID" parameter.

Example 6-6. Shift Register Class

```
#ifndef __SHIFT_CLASS__
#define __SHIFT_CLASS__

template<typename dataType, int NUM_REGS>
  class shift_class{
 private:
  dataType regs[NUM_REGS];
  bool en;
  bool sync_rst;
  bool ld;
  dataType *load_data;
 public:
    shift_class():en(true),sync_rst(false),ld(false){}
    shift_class(dataType din[NUM_REGS]):
      en(true),sync_rst(false),ld(false){
      load_data = din;
    }
    void set_sync_rst(bool srst){
      sync_rst = srst;
    }
    void load(bool load_in){
      ld = load_in;
    }
    void set_enable(bool enable){
      en = enable;
    }
    void operator << (dataType din){
    SHIFT:for(int i=NUM_REGS-1;i>=0;i--){
        if(en)
          if(sync_rst)
            regs[i] = 0;
          else if(ld)
            regs[i] = load_data[i];
          else
            if(i==0)
              regs[i] =  din;
            else
              regs[i] =  regs[i-1];
      }
    }
    dataType operator [](int i){
      return regs[i];
    }
};
#endif
```

Example 6-6 shows the implementation of the shift register class. The class implements all of the shift registers discussed so far with any combination of control signals. The implementation details are as follows:

- Line 4 is the template declaration with two template parameters, "dataType" and "NUM_REGS"
- Line 7 is the declaration of the "regs" array of type dataType and NUM_REGS elements
- Lines 8, 9, and 10 are internal control variables used for data enable, sync reset, and data load
- Line 11 is an internal pointer that is used to access the load data when a synchronous load is enabled
- Line 13 is the default constructor. This initializes the control variables "en", "srst" and "ld" so that the shift register is enabled, not reset, and not loading. These default values are constant propagated to remove the control logic when not used.
- Line 14 is the constructor that initializes the control variables and points the "load_data" variable to an array
- Lines 18, 19, and 24 are the member functions used to set the control variables. If these functions are not called then the control values always have the values assigned by the default constructor, and are constant propagated. This is how the different hardware configurations can be selected. Control logic is only inserted when a member function is used.
- Line 27 is the overloaded shift operator "<<" used to shift data though the "regs" array. This operator takes a right hand argument of type "dataType" and operates on "regs" based on the control variables. If a control variable is unused it is optimized away and does not cost any additional area. The order of operations are enable, sync reset, load, shift.
- Line 41 is the overloaded bracket operator "[]" which is used to index the "regs" array.

The shift register class of Example 6-6 now allows multiple shift registers of any arbitrary number of elements and data type. Furthermore each shift register instance can be configured to use any number of the control signals as needed. Unused control is optimized away. Example 6-7 shows a design that cascades two instances of the shift register class.

Example 6-7. Using the Shift Register Class

```
1  #include "test_shift_class.h"
2  #include "shift_class.h"
3  void shift_reg(dType din, dType load_data[N_REGS0],dType
   dout0[N_REGS0], dType dout1[N_REGS1], bool srst, bool load,
   bool en){
4    static shift_class<dType,N_REGS0> shift_reg0(load_data);
5    static shift_class<dType,N_REGS1> shift_reg1;
6
7    shift_reg0.set_enable(en);
8    shift_reg0.set_sync_rst(srst);
9    shift_reg0.load(load);
10   shift_reg1.set_enable(en);
11
12   shift_reg0 << din;
13   shift_reg1 << shift_reg0[N_REGS0-1];
14
15  WRITE0:for(int i=0;i<N_REGS0;i++)
16     dout0[i] = shift_reg0[i];
17  WRITE1:for(int i=0;i<N_REGS1;i++)
18     dout1[i] = shift_reg1[i];
19 }
20
```

```
Design Constraints
Main loop pipelined with II=1
All regs array mapped to registers
All loops fully unrolled
```

The details of Example 6-7 are:

- Lines 4 and 5 create two static instances of the shift register class. The instances are declared static so that the data inside of the class variable persists between function calls. The instances each have the same data type but N_REGS0 and N_REGS1 number of registers. The "shift_reg0" instance uses the constructor to point to "load_data" while "shift_reg1 does not support loading data and uses the default constructor.
- Lines 7 through 10 are used to set the control signals for each shift register. "shift_reg0" is configured to use all of the control signals built into the class, while "shift_reg1" only uses the "en" control signal.
- Lines 12 and 13 call the shift operator "<<" for both shift registers. "din" is shifted into "shift_reg0" and the right-most tap of "shift_reg0" is shifted into "shift_reg1"
- Lines 15 through 18 use the bracket operator "[]" to copy the shift register data to the outputs

Examples 6-6 and 6-7 begin to show the true power of C++ synthesis. Not only can any arbitrary length shift register be created from an instance of the shift register class, but any data type can be used, including complex user created classes provided that they implement the "<<" and "[]" operators.

Helper Classes for Design Reuse

The previous section of shift registers showed how to build reusable hardware by creating templatized C++ classes. This approach allowed the shift register to be parameterized based on the data type and number of registers. In that example there was a one to one correspondence between template parameters and the resulting hardware. In other words, the number of registers "NUM_REGS" in Example 6-7 was used directly in the underlying design to specify the number of array elements in "regs". However it is often desirable to statically compute some other internal parameter based on the template parameter. A good example of this is to compute the number of address bits required to index an array. There are a number of ways to compute these types of constants that leverage the power of C++ templates, including template recursion. Unfortunately these more powerful methods sometimes obscure the functionality within this type of template "magic". This section presents a more "brute force", but identical, approach to computing internal parameters. Template recursion is discussed in later sections.

Note

The helper classes described in this chapter are also supported by the Algorithmic C bit accurate data types.

Log2Ceil

One of the most commonly needed parameters is the number of bits required to index an array with N elements, or to count to N-1. This is known as the log2ceil function in C++ where it returns the value X that satisfies the condition N <= 2^X. Since this parameter N is usually based on a template parameter it requires the use of enumerated types to perform the computation so that the result is statically determinable at compile time.

An enumerated type is a set of named values where each value, known as an enumerator, usually behave as constants. A "helper class" is created to contain the enumerated type. This class is then used to compute the parameter. Example 6-8 shows the helper class for computing log2ceil up to 32 bits.

Example 6-8. Log2Ceil Helper Class

```
#ifndef __LOG2CEIL__
#define __LOG2CEIL__

template<int N_VAL>
struct LOG2_CEIL{
  enum {
    val = N_VAL <= 1 ? 1: N_VAL <= 2 ? 1 : N_VAL <= 4 ? 2 :
   N_VAL <= 8 ? 3 : N_VAL <= 16 ? 4 : N_VAL <= 32 ? 5 :
   N_VAL <= 64 ? 6 : N_VAL <= 128 ? 7 : N_VAL <= 256 ? 8 :
   N_VAL <= 512 ? 9 : N_VAL <= 1024 ? 10 : N_VAL <= 2048 ? 11 :
   N_VAL <= 4096 ? 12 : N_VAL <= 8192 ? 13 : N_VAL <= 16384 ? 14
:
   N_VAL <= 32768 ? 15 : N_VAL <= 65536 ? 16 : 32
  };
};

#endif
```

The enumerated type uses the template argument "N_VAL" and compares it to be less than or equal to 2^X. It returns X if true, otherwise it moves to the next comparison. The helper class can then be used directly to set the number of bits in a bit accurate data type. e.g.

```
ac_int<LOG2_CEIL<N_REGS>::val,false> addr;
```

One important point to note about using the "brute force" approach is that there must be sufficient enumerations to cover all of the possible values. This helper class is used in later sections.

NextPow2

Another constant that often must be computed is the next power of two of a number N that satisfies X = 2^Y for 2^Y >= N. This can also be computed using the same technique of a helper class and enumerated type, shown in Example. 6-9.

Example 6-9. NextPow2 Helper Class

```
#ifndef __NEXTPOW2__
#define __NEXTPOW2__

template<int N_VAL>
struct NEXT_POW2{
  enum {
    val = N_VAL <= 1 ? 1: N_VAL <= 2 ? 2 : N_VAL <= 4 ? 4 :
   N_VAL <= 8 ? 8 : N_VAL <= 16 ? 16 : N_VAL <= 32 ? 32 :
   N_VAL <= 64 ? 64 :  N_VAL <= 128 ? 128 :  N_VAL <= 256 ? 256
:
   N_VAL <= 512 ? 512 :  N_VAL <= 1024 ? 1024 :N_VAL <= 2048 ?
2048 :
   N_VAL <= 4096 ? 4096 : N_VAL <= 8192 ? 8192 : N_VAL <= 16384
? 16384 :
   N_VAL <= 32768 ? 32768 : N_VAL <= 65536 ? 65536 : 1048576
  };
};

#endif
```

Multiplexors

Two types of multiplexors are used during the HLS process, binary selection and onehot muxes. Binary selection MUXes are typically seen when performing simple indexing into an array mapped to registers, whereas onehot muxes are usually inferred when the indexing or control logic becomes more complicated. What usually determines the choice of MUX depends on the choice of C++ selection statement, the number of levels of MUXes, and the number of assignments involved in accessing the array. HLS typically optimizes multiple levels of binary selection muxes with common inputs into a single onehot MUX.

Binary MUX

The simplest, and most reliable, way to infer a two input binary section much is to use the question mark operator "?". Example 6-10 shows how the question mark operator is used to multiplex between two value.

Example 6-10. Two-to-one MUX Using the ? Operator

```
#include "binary_2x1_mux.h"
dType binary_2x1_mux(dType din[2],bool sel){
  return sel ? din[0]:din[1];
}
```

If a binary section MUX with more than two inputs is needed the array should be indexed with the selection variable. However care must be taken, not only to set the appropriate number of bits for the section variable, but also to limit the number of assignments. Example 6-11 shows how a single index into an array infers a binary selection MUX. With a binary section MUX the number of bits of the selection variable should be log2ceil of the number of MUX elements, which in this example are equal to “N_REGS”. The log2ceil helper class covered in “Helper Classes for Design Reuse” on page 123 can be used to compute the proper number of bits for “sType”. This is done in the “binary_mux.h” include file shown in Example 6-12.

Example 6-11. Binary Selection MUX

```
#include “binary_mux.h”
dType binary_mux(dType din[N_REGS],sType sel){
  return din[sel];
}
```

Example 6-12. Binary Selection MUX Header File

```
1   #ifndef __BINARY_MUX__
2   #define __BINARY_MUX__
3   #include <ac_int.h>
4   #include “../../helper_classes/src/log2ceil.h”
5   #define N_REGS 8
6
7   typedef ac_int<8> dType;
8   typedef ac_int<LOG2_CEIL<N_REGS>::val,false> sType;
9   dType binary_mux(dType din[N_REGS],sType sel);
10
11  #endif
12
```

Line 8 of Example 6-12 shows how “sType” is defined with the minimum number of bits required to index an array with N_REGS elements. The LOG2CEIL helper class is used to statically compute log2ceil of “N_REGS”.

Automatic Binary to Onehot MUX Optimizations

HLS automatically optimizes cascaded binary MUXes into onehot MUXes if common inputs exist in the cascade structure. This optimization in general gives better performance at the cost of slightly larger area. A common scenario where this occurs is shown in Example 6-13.

Example 6-13. Automatic MUX Optimizations

```
#include "binary_mux.h"
dType binary_mux(dType din[N_REGS],sType sel0, sType sel1, bool
s){
    dType tmp;
    if(s)
        tmp = din[sel0];
    else
        tmp = din[sel1];
  return tmp;
}
```

The general hardware structure of Example 6-13 before and after optimizations is shown in Figure 6-5. The MUX tree is optimized into a single onehot MUX since "din" is common to both MUXes on the input. The un-optimized hardware shows that the two branches of the "if" statement controlled by "s" become inputs into a 2-to-1 MUX. Each of the "if" branches are fed with a binary section MUX controlled by "sel0" and "sel1".

Figure 6-5. Automatic MUX Optimizations

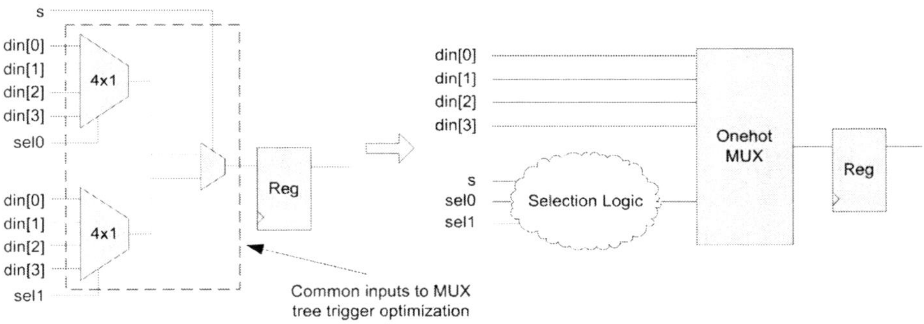

Manual Optimization of Binary Selection MUXes

Example 6-13 showed that multi-level binary selection MUX structures can be automatically optimized into onehot MUXes. The C++ description must be restructured if the desired behavior is a single binary selection MUX to select between elements of "din". The key is to reduce the accesses of "din" to a single point in the C++ code and to explicitly code the selection logic, shown in Example 6-14. The control variable "s", line 5, is now used to select between "sel0" and "sel1" and assign to an internal variable "sel_int". "sel_int" is then used to access "din", line 9, in a single location.

Example 6-14. Manual Optimization of MUXes

```
#include "binary_mux.h"
dType binary_mux(dType din[N_REGS],sType sel0, sType sel1,
bool s){
   dType tmp;
   sType sel_int;
   if(s)
     sel_int = sel0;
   else
     sel_int = sel1;
   return din[sel_int];
}
```

The hardware diagram for Example 6-14 is shown in Figure 6-6.

Figure 6-6. Manual Optimization of MUXes

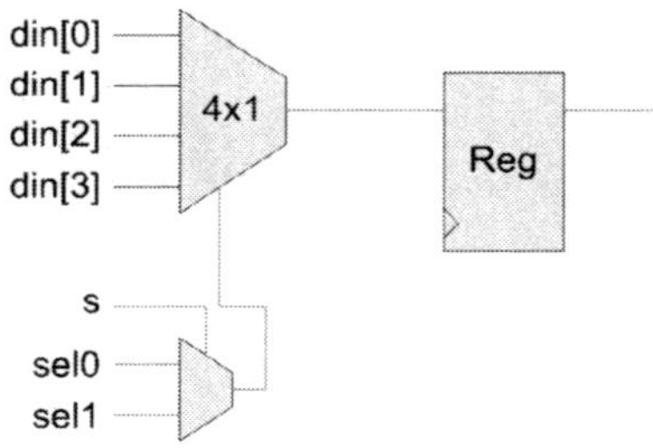

One Hot MUX

In addition to the automatic onehot MUX optimizations discussed previously, it is also possible to explicitly code onehot MUXes using the "switch" or "if-else" statements. Example 6-15 shows the use of a "switch" statement that is inferred as a onehot MUX. In this example HLS encodes the selection logic to prevent multiple selections at the same time. In many cases the onehot MUX is controlled by the data path FSM which is onehot encoded.

Example 6-15. Onehot MUX Using "switch" Statements

```
#include "onehot_mux.h"
dType onehot_mux(dType din[N_REGS],sType sel){
  dType tmp;
  switch(sel){
  case 1: tmp = din[0];
    break;
```

```
    case 2: tmp = din[1];
      break;
    case 4: tmp = din[2];
      break;
    case 8: tmp = din[3];
      break;
    default: tmp = 0;
      break;
    }
    return tmp;
}
```

Example 6-16 shows the use of an “if-else” statement that causes a onehot MUX to be inferred.

Example 6-16. Onehot MUX using “if-else” Statements

```
#include "onehot_mux.h"
dType onehot_mux(dType din[N_REGS],sType sel){
  dType tmp;
  if(sel==1)
     tmp = din[0];
  else if(sel==2)
    tmp = din[1];
  else if(sel==4)
    tmp = din[2];
  else if(sel==8)
    tmp = din[3];
  else
    tmp = 0;
  return tmp;
}
```

Priority Search Hardware

One of the more common functions encountered in many designs is some form of a priority search such as finding the position of the first leading one in a bit-vector, or finding the minimum or maximum value in an array. Although these types of algorithms are very easy to express in C++ using a for loop and some counters or comparators, the resulting hardware is not always optimal. It can be on the order of N levels of logic, where N is the size of the search. Many of these algorithms can be realized in log2(N) levels of logic when written slightly differently.

Finding Leading 1's in a Bit-vector

Algorithmic Coding Style

Example 6-17 shows the most common way to code an algorithm that returns the position of the first leading one in the bit-vector as well as a flag that indicates if any or none of the bits are set.

Example 6-17. Finding Leading Ones in a Bit-vector

```
1   #include "find_leading_ones.h"
2   bool find_leading_ones(ac_int<NUM_BITS,false> din,
3                                             ac_int<LOG2_CEIL<NUM_
BITS>::val,0> &dout){
4     int tmp;
5     bool flag = false;
6     for(int i=NUM_BITS-1;i>=0;i--){
7       if(din[i]){
8         flag = true;
9         tmp = i;
10        break;
11      }
12    }
13    dout = tmp;
14    return flag;
15  }
16

Design Constraints
Main loop pipelined with II=1
All loops fully unrolled
NUM_BITS = 32
```

The details of Example 6-17 are:

- Line 3 - "dout" returns the bit position of the first leading one. This means that the maximum count can be represented with log2ceil(NUM_BITS) bits. The helper class covered in "Helper Classes for Design Reuse" on page 123 is used to compute the minimum number of bits for "dout".
- Line 5 - the "flag" variable is initialized to false. If not set it is returned indicating that there are no ones in the bit vector.
- Lines 6 though 11 - Starting with the uppermost bit each bit is checked and the loop is exited if a bit is set to one.

The hardware diagram for Example 6-17 is shown in Figure 6-7. The bits of “din” are used to generate the section logic for a 32x1 onehot MUX, which has the position count as its data inputs.

Figure 6-7. Finding Leading Ones in a Bit-vector

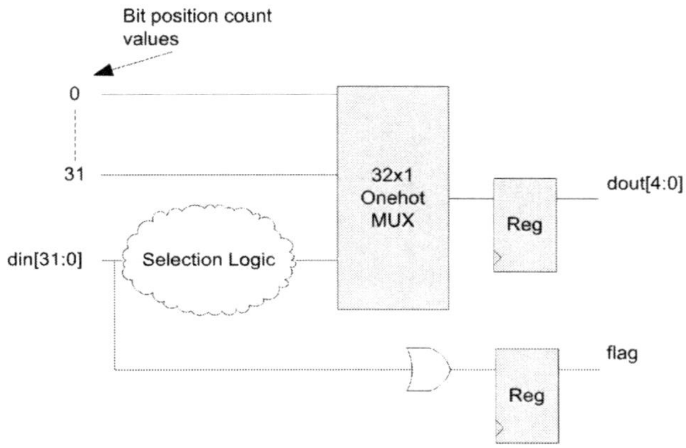

Improved Performance and Area Using the Brute Force Approach

Although there are much more efficient ways to code Example 6-17, they require considerable more thought into what the underlying hardware should look like. Sometimes it is only necessary to improve performance and area by a little bit in order to hit the desired metrics. In these cases it is often easier to use a more brute force approach to subdivide the algorithm implementation into smaller chunks. Example 6-18 shows a rewrite of the leading ones algorithm that divides the original algorithm into two parts.

Example 6-18. Finding Leading Ones Using Brute Force

```
#include <ac_int.h>
#include "find_leading_ones.h"
bool  find_leading_ones(ac_int<NUM_BITS,false>  din,
ac_int<LOG2_CEIL<NUM_BITS>::val,0> &dout){
  int upper=0,lower=0;
  bool flagu = false;
  bool flagl = false;
  for(int i=NUM_BITS-1;i>=NUM_BITS/2;i--)
    if(din[i]){
      upper = i;
      flagu = true;
      break;
    }
  for(int i=NUM_BITS/2-1;i>=0;i--)
```

```
14      if(din[i]){
15        lower = i;
16        flagl = true;
17        break;
18      }
19    dout =  flagu ? upper:lower;
20    return flagu|flagl;
21  }

Design Constraints
Main loop pipelined with II=1
All loops fully unrolled
NUM_BITS = 32
```

The details of Example 6-18 are:

- Line 4 - two counters used to store the bit position of the leading one for upper and lower halfs of the bit vector
- Lines 5 and 6 - flags for both the upper and lower half of the bit-vector
- Lines 7 through 18 - two loops are used to look for leading ones in the upper and lower half of the bit-vector
- Line 19 - if any upper bit is set return the upper count, otherwise return the lower count
- Line 20 - “or” the upper and lower flags and return

The hardware diagram of Example 6-18 is shown in Figure 6-8. The onehot MUX has been reduced from a 32x1 to a 16x1 and the MSB of the position count is set based on the upper flag “flagu”. This is a result of the search being performed on a 32-bit vector. If the vector was not a power of two the logic would be slightly more complex. However, as shown in the next few sections, it is possible to zero pad the input bit-vector to make it a power of two. This “brute force” approach can be used to further divide the problem into smaller chunks, but there are more elegant ways to do this.

Figure 6-8. Finding Leading One Using Brute Force

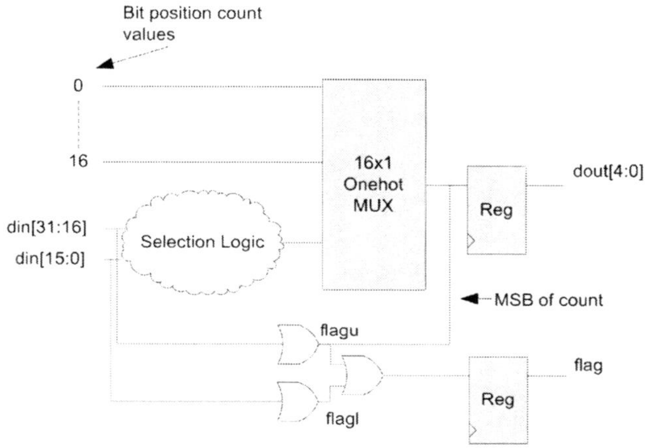

Log2(N) Based Search

The optimal algorithm for finding the leading ones in an N-bit bit-vector should take log2ceil(N) iterations to complete. This algorithm is similar to the "brute force" approach, but it continues dividing the vector into upper and lower parts until it operates on a single bit. Example 6-19 shows this implementation.

Example 6-19. Finding Leading Ones Using Log2(N) Search

```
#include "find_leading_ones.h"
#include <ac_int.h>
bool find_leading_ones(ac_int<NUM_BITS,false> din,
                       ac_int<LOG2_CEIL<NUM_BITS>::val,0> &dout){
  enum {P2 = NEXT_POW2<NUM_BITS>::val};
  enum {L2 = LOG2_CEIL<NUM_BITS>::val};
  int tmp;
  ac_int<P2,false> upper,lower;
  ac_int<P2,false> mask = 0;
  ac_int<P2,false> din_tmp=0;
  bool flag = false;
  int idx = 0;
  din_tmp = din;
  mask = ~mask;
  flag = din_tmp?1:0;
  for(int i=0;i<L2;i++){
    mask = mask >> ((P2/2)>>i);
    upper = lower = 0;
    upper = din_tmp>>((P2/2)>>(i));
```

```
20        lower = din_tmp&mask;
21        din_tmp = 0;
22        if(upper){
23          idx = idx + (P2/2 >> i);
24          din_tmp = upper;
25        }else
26          din_tmp = lower;
27      }
28      dout = idx;
29
30      return flag ;
31  }
```

```
Design Constraints
Main loop pipelined with II=1
All loops fully unrolled
NUM_BITS = 32
```

The details of Example 6-19 are:

- Lines 4, 5, and 6 - use the helper functions for computing log2ceil(N) and nextpow2(N). Enumerated types are used to compute these values, "P2" and "L2",since they are used in multiple locations within the design.
- Lines 8, 9, and 10 - define internal variables as power of two bits wide. This allows support for any N bits by zero padding the internal variable "din_tmp".
- Lines 12, 13, and 14 - clear the position count, copy the input and zero pad if necessary, and set all mask bits equal to one.
- Line 15 - set the flag if any bit in the bit vector is set to one.
- Line 16 - iterate on "din_tmp" log2ceil(N) times.
- Lines 17 through 20 - Half the number of mask bits for each successive iteration and mask off the upper and lower portions of the bit-vector.
- Lines 22 though 27 - Check to see if a one is set in "upper", if set add P2/2>>i to the position count, and set "din_tmp" equal to the upper half. Otherwise set "din_tmp" equal to the lower half and go to the next iteration until finished.

The general hardware structure of Example 6-19 is shown in Figure 6-9 with some gates omitted for clarity. What is shown is that the upper sections are "OR'd" together and the output of the "or" gate sets a bit of the position count "dout" and selects the next upper or lower section.

Figure 6-9. Finding Leading Ones Using Log2(N) Search

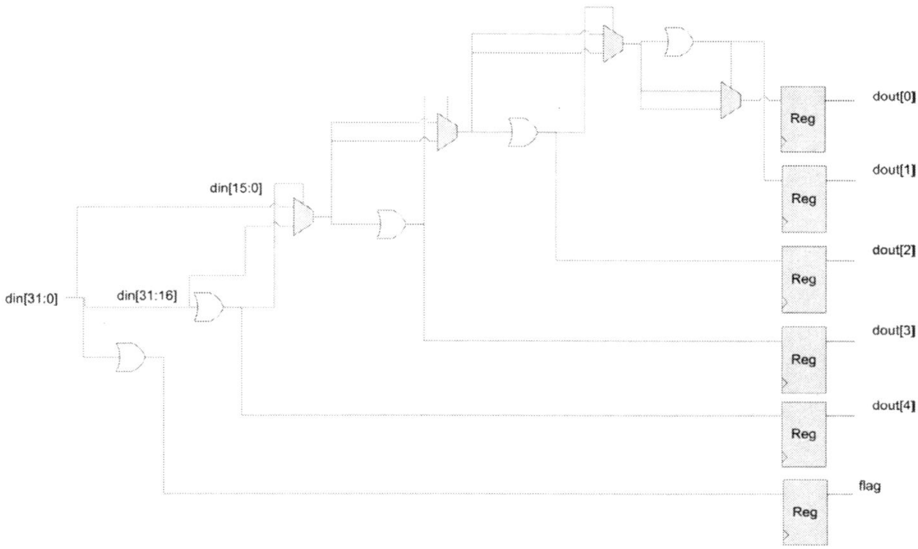

Recursive Template Search

The previous examples on finding the leading one in a bit-vector showed that more efficient hardware can be realized by coding more hardware intent into the C++. The log2 based search provides optimal hardware, but there is an alternative implementation that can provided similar results using C++ template recursion. Template recursion has the advantage of allowing designers to build highly balanced hardware. One drawback of using recursive templates is that there is no capability for design exploration via loop unrolling because the design is fully parallel. However algorithms such as finding the leading ones or the maximum value of an array are often fully parallel. Example 6-20 shows a recursive template function implementing of finding the leading ones in a bit vector.

Example 6-20. Find Leading Ones Using Recursive Template Search

```
#ifndef __LEADING_ONES__
#define __LEADING_ONES__
#include <ac_int.h>
#include "../../helper_classes/src/log2ceil.h"
#include "../../helper_classes/src/nextpow2.h"
template<int N_BITS>
```

```
7   bool leading_ones(ac_int<N_BITS,false> &din,
8                               ac_int<LOG2_CEIL<N_BITS>::val,false>
&dout){
9     enum {
10      P2 = NEXT_POW2<(N_BITS+1)/2>::val
11    };
12    ac_int<N_BITS-P2,false> upper;
13    ac_int<P2,false> lower;
14    ac_int<LOG2_CEIL<N_BITS>::val,0> idx=0;
15    ac_int<LOG2_CEIL<N_BITS-P2>::val,0> idxu=0;
16    ac_int<LOG2_CEIL<P2>::val,0> idxl=0;
17    static bool flag = false;
18
19    upper.set_slc(0, din.template slc<N_BITS-P2>(P2));
20    lower.set_slc(0, din.template slc<P2>(0));
21
22    if(upper){
23      leading_ones<N_BITS-P2>(upper,idxu);
24      idx = idxu | P2;
25    }
26    else{
27      leading_ones<P2>(lower,idxl);
28      idx = idxl;
29    }
30    dout = idx;
31    return flag = (din!=0) ?1:0;
32  }
33
34  template<>
35  bool leading_ones<1>(ac_int<1,false> &din,
36                           ac_int<1,false> &dout){
37    dout = 0;
38    return din[0];
39  }
40  #endif
```

The details of Example 6-20 are:

- Lines 8 and 10 use the helper classes for computing the log2ceil(N_BITS) and nextpow2(N_BITS+1). The nextpow2 computation allows the design to handle bit-vectors that are not a power of two. The reason why we use nextpow2(N_BITS+1) is to ensure that vectors with an odd number of bits split the bits so that the upper half is a power of two.
- Line 12 declares the number of bits for the upper half of the bit-vector. This takes into account if the bit-vector is not a power of two by using N_BITS-P2.
- Line 13 declares the number of bits for the lower half of the bit vector. This is always a power of two, P2 bits.
- Lines 14, 15, and 16 define the current index, and the index variables for the upper and lower halfs of the bit-vector.

- Lines 19 and 20 slice the bit vector into the "upper" and "lower" variables.
- Lines 22 through 25 check to see if any of the bits in "upper" are set, and if true, recursively calls the "leading_ones" function, passing it all of the "upper" bits. Since the upper half of the bit-vector is chosen, the bit index offset, P2, is added to the previous index "idxu".
- Lines 29 through 29 pass the "lower" variable to the recursive call of "leading_ones" and sets the current index to the previous index "idxl".
- Line 31 sets flag to true if any bit is equal to one.
- Lines 34 to 39 implement the specialization of the "leading_ones" function for N_BITS==1.

Finding the Maximum Value in an Array

Another commonly used function, or algorithm, in hardware design is to determine the maximum or minimum value in a sequence of values. In C++ this is typically done by searching the elements of an array for the maximum or minimum. The implementation of this in synthesizable C++, and ultimately the quality of results, has many similarities to the algorithm for finding the leading one in a bit vector, discussed in "Finding Leading 1's in a Bit-vector" on page 129.

Algorithmic Coding Style

Similar to finding the leading one in a bit-vector, the max search algorithm can be expressed very compactly in C++. Example 6-21 shows the C++ implementation for searching an array on integers.

Example 6-21. Finding the Maximum Value in an Array

```
#include "test_max.h"
void test_max(int din[N_REGS], int &dout){
  int max;
  int tmp;

  for(int i=0;i<N_REGS;i++){
    if(i==0)
      max = din[i];
    else{
      tmp = din[i];
      if(tmp>max)
```

```
12          max = tmp;
13       }
14    }
15    dout = max;
16 }
17

Design Constraints
Main loop pipelined with II=1
"din" mapped to registers or wire interface
All loops fully unrolled
N_REGS = 4
```

The details of Example 6-21 are:

- Line 8 reads the first element of "din" and assigns it to "max". This is to avoid making one extra comparison.
- Lines 11and 12 check to see if "tmp" is greater than the previous "din[i]" and keeps the larger of the two.

The C++ implementation of Example 6-21 is very clear and compact. However the quality of results in terms of area and performance may be less than ideal in some situations. This coding style is generally acceptable when the loop is left rolled and/or "din" is mapped to a memory. However, when "din" is mapped to registers and the loop is fully unrolled the resulting hardware has on the order of "N_REGS" levels of logic. This is because this type of C++ description is a "priority" encoded type structure. In other words the bigger the array gets, the longer the delay of the algorithm. This can have unwanted consequences such as more registers than required if the design is scheduled across multiple clock cycles, or worst case failure to schedule if the design sits in a pipeline feedback path. Figure 6-10 shows the approximate schedule for Example 6-21. The priority encoded nature of this algorithm is more likely to require multiple clock cycles to schedule as the clock frequency is increased. This is usually undesired behavior because it means larger area.

Figure 6-10. Schedule of Priority Encoded Search

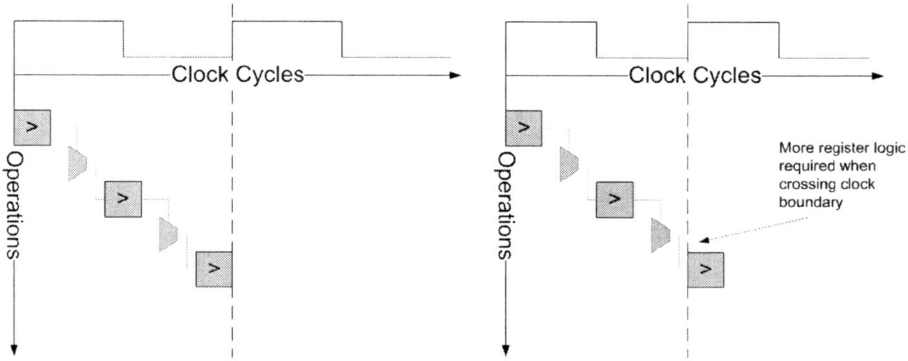

The hardware diagram for Example 6-21 is shown in Figure 6-11 assuming it has been scheduled within a single clock cycle. This shows the "serial" nature of the comparisons, which is essentially what the C++ describes. Although this implementation may be more than adequate for the end application, there are alternative ways to code this algorithm to achieve better performance and area. One approach would be to use the "brute-force" approach covered in "Improved Performance and Area Using the Brute Force Approach" on page 130. This approach would manually subdivide the comparisons into separate halfs. The most optimal solution for this type of algorithm is to use a recursive template function to fully subdivide the problem, which leads to a balanced comparison tree.

Figure 6-11. Hardware of Priority Encoded Search

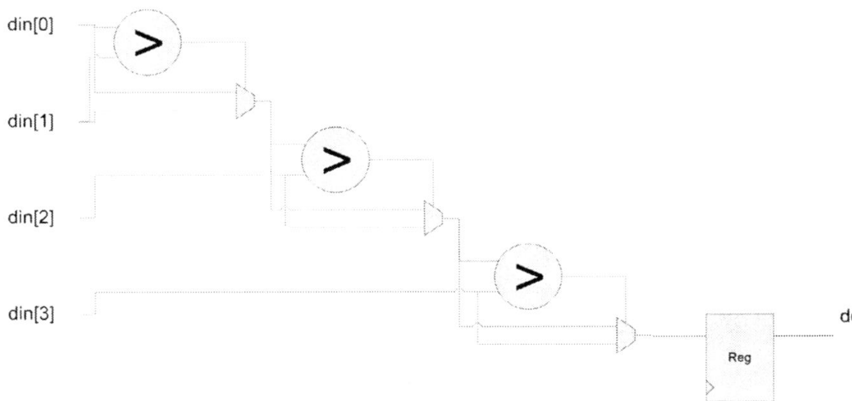

Recursive Template Search

Unlike the "finding leading ones" algorithm in "Finding Leading 1's in a Bit-vector" on page 129, which can be implemented optimally using either recursive template functions or loops, the max search algorithm cannot be written optimally using loops. Using recursive template functions allows one to realize the most efficient hardware for both area and performance by building a balanced structure that has on the order of log2(N) levels of logic, with N being the number of array elements. One issue with recursive template functions is that partial specialization is not supported, which poses a problem since it is desirable to specify both the data type and the number of elements as template parameters. To make the max search algorithm truly generic a "helper" struct can be used to work around the limitations of partial specialization [1].

Example 6-22 shows the recursive template implementation of the max search algorithm.

Example 6-22. Max Search Using Recursive Templates

```
// helper struct
template<int N>
struct max_s {
  template<typename T>
  static T max(T *a) {
    T m0 = max_s<N/2>::max(a);
    T m1 = max_s<N-N/2>::max(a + N/2);
    return m0 > m1 ? m0 : m1;
  }
};
// terminate template recursion
template<> struct max_s<1> {
  template<typename T>
  static T max(T *a) {
    return a[0];
  }
};
template<int N, typename T>
  T max(T *a) {
  return max_s<N>::max(a);
}

```

The details of Example 6-22 are:

- Lines 18 and 19 show a templatized function "max" that takes the number of array elements and the data type as the template arguments.
- Line 20 uses the helper struct "max_s" to work around partial specialization. The number of elements N are passed as the only struct template parameter, and template substitution is used to know what the data type of "a" is when calling the "max" member function of "max_s".
- Lines 2 and 3 define the templatized helper struct. The helper struct has only N, the number of array elements, as its template argument.
- Lines 4 and 5 define the helper struct "max" function which has the data type "T" as its template argument. A pointer to the array to be searched is passed to this function.
- Lines 6 and 7 recursively call the helper struct "max" function, dividing the array into upper and lower halfs. Each recursive function call returns a comparison value "m0" and "m1".
- Line 8 compares "m0" and "m1" and returns the maximum value.
- Lines 12 through 16 implement the specialization for "max_s". This function returns the array elements themselves.

Example 6-23 shows the "max" function used in a top-level design.

Example 6-23. Instantiating the Recursive Template Function

```
#include "test_max.h"
#include "max.h"
void test_max(int din[N_REGS], int &dout){
  dout = max<N_REGS>(din);
}

Design Constraints
Main loop pipelined with II=1
"din" mapped to registers or wire interface
N_REGS = 4
```

Figure 6-12 shows the data flow graph for recursive template implementation of the max function.

Figure 6-12. Data Flow Graph of Recursive Template Max Function

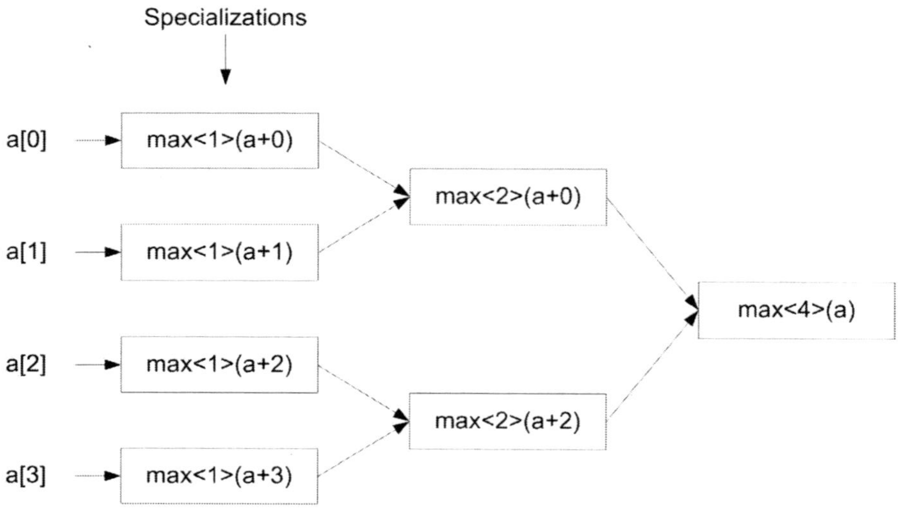

Figure 6-13 shows the hardware synthesized from Example 6-23. The result is a balanced comparison tree that yields the best area and performance.

Figure 6-13. Hardware Implementation of Recursive Max Function

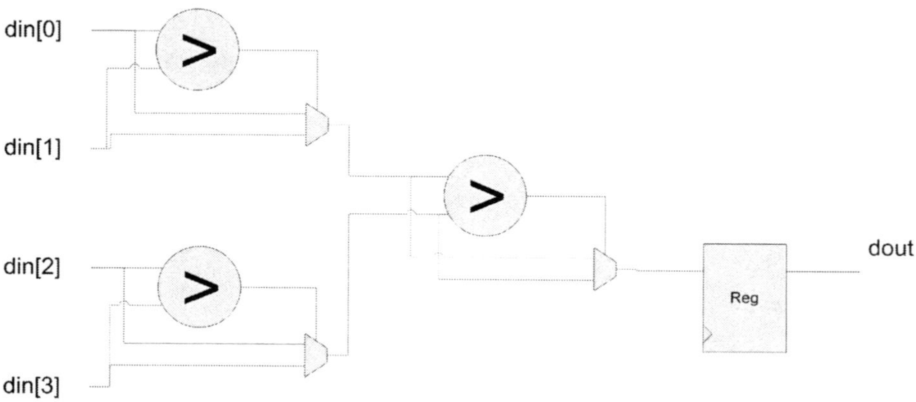

Absolute Value (abs)

Calculating the absolute value of a number is a function that is often used in mathematics and many DSP algorithms. It is usually expressed as:

$$a = \begin{cases} a & a \quad 0 \\ -a & a < 0 \end{cases}$$

In most cases it is sufficient to express this algorithmically, shown below in Example 6-24.

Example 6-24. Absolute Value

```
#include <ac_int.h>
ac_int<8> abs(ac_int<8> din){
  ac_int<8> tmp = din;
  if(tmp<0)
    tmp = -tmp;
  return tmp;
}
```

```
Design Constraints
Main loop pipelined with II=1
```

The hardware diagram for Example 6-24 is shown below in Figure 6-14.

Figure 6-14. Hardware of Absolute Value

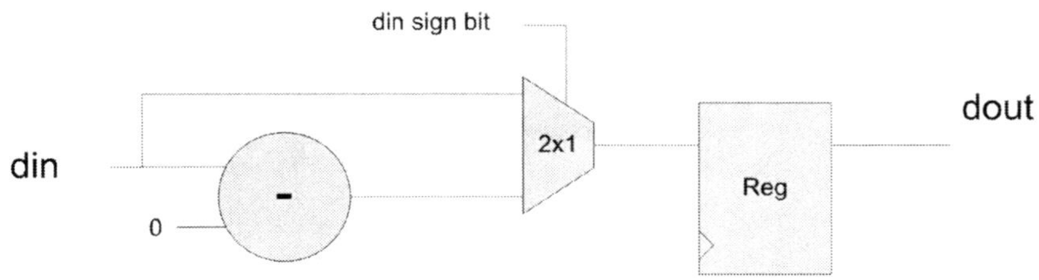

The hardware that gets synthesized uses the sign bit of "din" to select between "din" and "-din". If area must be reduced in a design it's possible to re-code the "abs" function using bit-level expressions which can help reduce area by as much as 10 to 20 percent in some instances. This area improvement may be negligible in the context of a much larger design, but the cumulative effect of making these code transformations throughout a design can be substantial. Example 6-25 shows the "abs" function rewritten using bit-level expressions that eliminate the need for a MUX, instead using XOR gates which require less area.

Example 6-25. Bit-level Implementation of Absolute Value

```
1   #include <ac_int.h>
2   ac_int<8> abs(ac_int<8> din){
3     ac_int<8> tmp0=0,tmp1 = 0;
4     tmp0 = din;
5     for(int i=0;i<8;i++)
6       tmp1[i] = tmp0[i]^tmp0[7];
7     return tmp1+tmp0[7];
8   }
```

```
Design Constraints
Main loop pipelined with II=1
All loops fully unrolled
```

The details of Example 6-25 are:

- Lines 1 through 3 show that this design uses bit-accurate data types. This makes it much easier to perform bit level operations using the ac_int and ac_fixed "[ ]" bit slice operator.
- Lines 5 and 6 XORs the sign bit of "din", "din[7]",with all the bits of "din" which has been copied into the "tmp0" variable. This essentially inverts "din" if the sign bit is set.
- Line 7 takes "tmp1" and adds the sign bit of "din". It can be seen that lines 5 through 7 have implemented a twos-complement negation when the sign bit is set, shown in Figure 6-15.

Figure 6-15. Hardware for Bit-level Implementation of Absolute Value

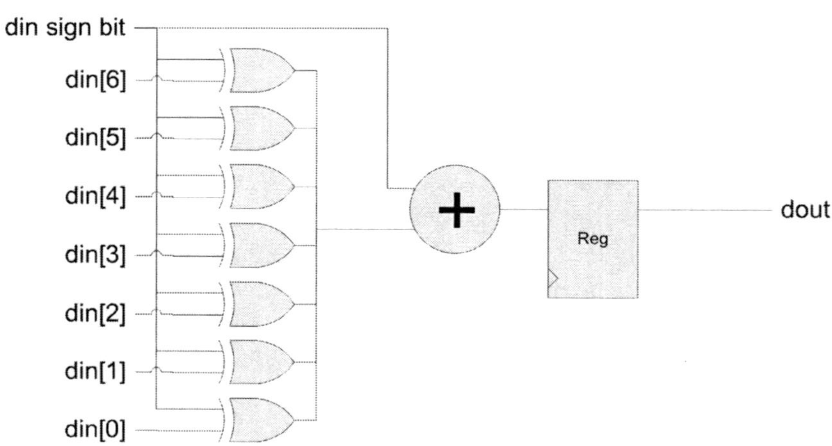

Although Example 6-25 requires a little more thought to express the absolute value for inputs of types "ac_int<8>, it can easily be generalized to support any size bit accurate integer data type by rewriting it as a template function, shown in Example 6-26.

Example 6-26. Generic Bit-level Implementation of Absolute Value

```
1   #ifndef __ABS_OPT_TEMPLATE__
2   #define __ABS_OPT_TEMPLATE__
3   #include <ac_int.h>
4
5   template<int NUM_BITS>
6   ac_int<NUM_BITS> abs(ac_int<NUM_BITS> din){
7     ac_int<NUM_BITS> tmp0=0,tmp1 = 0;
8     tmp0 = din;
9     for(int i=0;i<NUM_BITS;i++)
10      tmp1[i] = tmp0[i]^tmp0[NUM_BITS-1];
11    return tmp1+tmp0[NUM_BITS-1];
12  }
13  #endif
```

The details of Example 6-26 are:

- Line 5 - a single template parameter specifies the number of bits in the ac_int.
- Line 6 - this function is hard coded to use only signed ac_int data types.
- Lines 9 through 11 - the template parameter "NUM_BITS" is used extract the sign bit and control the loop iterations.

Linear Feedback Shift Register (LFSR)

An LFSR is a shift register whose input is a function of the state of some of the previous shift register bits. LFSRs are used in a wide range of applications ranging from Cryptography to communications. LFSRs are also very good for implementing very fast counters since the feedback in minimal compared to traditional binary counters. The feedback to the input of an LFSR can be represented as a mod-2 polynomial. E.g.

```
input bit = x3 + x2 + 1
```

Because the polynomial is mod-2 the input bit is equal to the XORing of the taps, excluding tap 0 which has no effect. Bit accurate data types make expressing an LFSR very easy. Example 6-27 shows a four tap loadable LFSR.

Example 6-27. Linear Feedback Shift Register

```
1   #include <ac_int.h>
2   void lfsr(ac_int<4,false> load_data, bool ld,
    ac_int<4,false> &dout){
3     static ac_int<4> reg;
4     ac_int<1,false> bit;
5
6     if(ld)
7       reg = load_data;
8     bit = reg[3] ^ reg[2];
9     reg<<=1;
10    reg[0] = bit;
11    dout = reg;
12  }
```

Design Constraints
Main loop pipelined with II=1

The details of Example 6-27 are:

- Lines 2 and 3 - unsigned bit-accurate data types are idea for implementing LFSRs. No loops are required.
- Line 6 and 7 - the LFSR is loaded when "ld" is true. NOTE that this implementation assumes that the LFSR is not loaded with zero. If zero is loaded the LFSR does not count.
- Line 8 implements the feedback polynomial.
- Line 9 shifts the bits of the LFSR
- Line 10 assigns the feedback polynomial result to bit zero of the LFSR

The hardware for Example 6-27 is shown below in Figure 6-16.

Figure 6-16. Hardware for Loadable LFSR

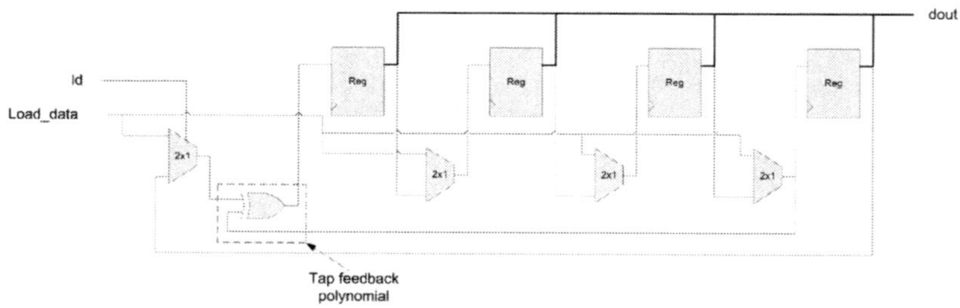

Accumulator

Accumulating the elements of an array is a common function seen in a variety of applications such as FIR filters, image processing, etc. The accumulation is typically accomplished using a loop to index the elements of the array. Fully unrolling the loop generally yields the optimal implementation of the sum of all array elements. This is because fully unrolling a loop allows bit widths of intermediate variables to be automatically reduced. However, care must be taken in writing the C++ when loops are left rolled, since automatic bit width reduction may not occur. The best coding practice is to account for the bit growth required for intermediate storage when accumulating an N element array. This bit growth is based on the bit-width of the input data type and the number of elements in the array. The number of extra bits needed to avoid overflow is log2ceil(N) bits. Bit-accurate data types allows the bit growth to be controlled explicitly. Example 6-28 shows a templatized implementation of an accumulator with bit growth computed based on the width of the data type as well as the number of elements. This implementation is designed to work with ac_int data types.

Example 6-28. Templatized Accumulator

```
1   #ifndef __ACCUM__
2   #define __ACCUM__
3   #include <ac_int.h>
4   #include "../../helper_classes/src/log2ceil.h"
5   template<int W, bool S, int N>
6     ac_int<W+LOG2_CEIL<N>::val,S> accumulate(ac_int<W,S> din[N])
{
7      ac_int<W+LOG2_CEIL<N>::val,S> acc = 0;
8
9    ACCUM:for(int i=0;i<N;i++){
10        acc += din[i];
11     }
12     return acc;
13  }
14  #endif
```

The details of Example 6-28 are:

- Line 5 specifies three template parameters, “W” for the bit width, “S” for the signedness, and “N” for the number of array elements.

- Lines 6 and 7 use the helper class LOG2_CEIL to compute the bit growth for the return type and the internal storage variable "acc".

Similarly to the other templatized functions discussed previously, the accumulator is used by instantiating it in another C++ design, and specifying the template parameters as shown in Example 6-29.

Example 6-29. Instantiating the Accumulator

```
#include "accumulate.h"
#include "test_accumulate.h"
void test_accumulate(ac_int<WIDTH> din[NUM_REGS],
                     ac_int<WIDTH_OUT> &dout){
  dout = accumulate<WIDTH,SIGN,NUM_REGS>(din);
}

Design Constraints
Main loop pipelined with II=1
All loops left rolled
WIDTH=8, SIGN=true, NUM_REGS=4, and WIDTH_OUT=10
```

The hardware for Example 6-29 is shown below in Figure

Figure 6-17. Hardware for Accumulator

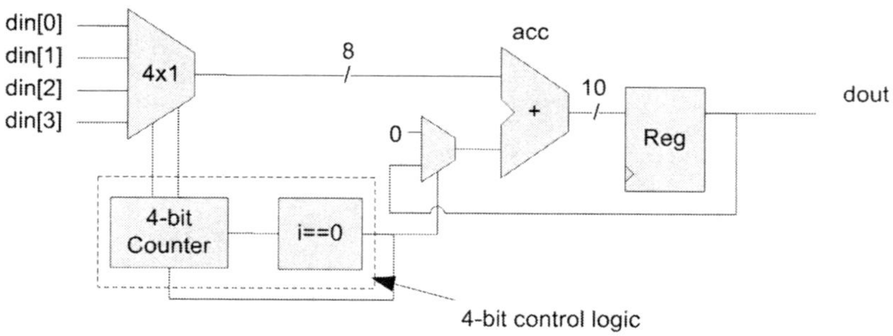

Shifters

The process of shifting a bit vector, either dynamically or statically, in C++ is easily expressed using the built-in shift operators "<<" and ">>". What is sometimes not as obvious is the resulting hardware based on the "signedness" and the bit widths of the arguments to the shift operators. As is usually the case with HLS, care should be taken when using the shift operator to ensure that the resulting hardware has the desired implementation.

Barrel shifter

A barrel shifter allows a bit-vector to be shifted by an arbitrary amount in a single clock cycle. The shift direction can either be left or right, and the shift can be logical or arithmetic. HLS has built-in operator support for barrel shifters, and uses them when scheduling a design as needed. An important point to remember is that barrel shifters can be costly in terms of area and performance, so understanding how they are inferred influences quality of results. The underlying hardware implementation of the full barrel shift operation consists of at most N*log2(N) 2-input multiplexors, where N is the number of bits in the bit-vector being shifted.

Logical

A logical barrel shift is inferred when the data variable being dynamically shifted is unsigned. Logical shifts insert zeros into the MSB or the LSB depending on the shift direction. Figure 6-18 shows how data is shifted in an 8-bit vector one bit position. Zero's are stuffed into the MSBs and the LSBs are discarded for logical shift right. Logical shift left stuffs zeros into the LSBs and discards the MSBs

Figure 6-18. Logical Shift Left and Right

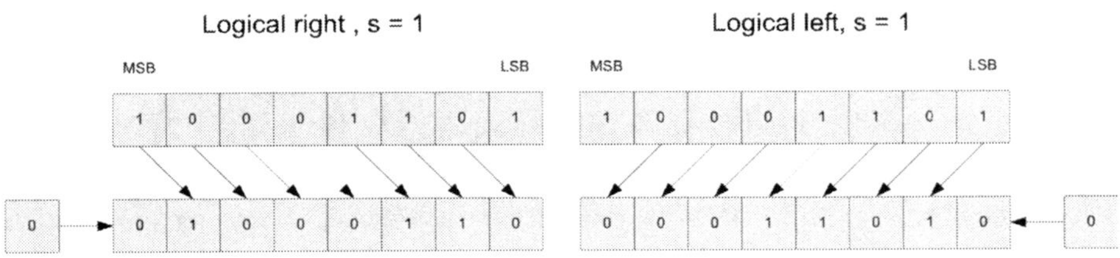

Example 6-30 shows a design that causes a logical shift right barrel shifter to be inferred and scheduled. The ac_int data types are used which allow the signedness to be expressed as a template parameter. In this case the data variable that is shifted is declared as unsigned, which causes a logical shifter to be inferred for the “>>” operator.

Example 6-30. Barrel Shifter with Logical Shift Right

```
#include "barrel_shift_lr.h"
ac_int<NUM_BITS,false> barrel_shift_lr(ac_int<NUM_BITS,false>
din,
                                                    ac_int<CTRL_
BITS,false> s){
  return din >> s;
}

Design Constraints
NUM_BITS = 8, CTRL_BITS = 4
```

The number of control bits needed to shift a vector of NUM_BITS is equal to log2ceil(NUM_BITS) + 1. The computation of CTRL_BITS can be done automatically by leveraging the helper class for log2ceil. Example 6-31 shows how the helper class can be used directly in the header file that defines the design parameters.

Example 6-31. Computing Barrel Shifter Control Bit Width

```
1   #ifndef __BARREL_SHIFT__
2   #define __BARREL_SHIFT__
3
4   #include <ac_int.h>
5   #include "../../helper_classes/src/log2ceil.h"
6   #define NUM_BITS 8
7   #define CTRL_BITS LOG2_CEIL<NUM_BITS>::val+1
8
9   ac_int<NUM_BITS,false> barrel_shift_lr(ac_int<NUM_BITS,false>
din,
10                                                        ac_
int<CTRL_BITS,false> s);
11  #endif
```

Lines 5 though 7 of Example 6-31 show another way that the LOG2CEIL helper class can be used to statically compute the bit widths of variables inside of a header file. It is important to make the width of the barrel shifter control variable only as wide as is needed. Otherwise there is an additional overhead in and/or logic to account for the upper bits of the shift control. Example 6-30 could be easily converted into a logical left shift barrel shifter by using the left shift operator "<<".

Arithmetic

An arithmetic barrel shifter is inferred when the data variable being dynamically shifted is signed. The arithmetic shift differs slightly from the logical shift in that zeros are only stuffed into the LSB when left shifting. The MSB, or sign bit is extended for right shifts, and is maintained for left shifts until overflow occurs. Figure 6-19 shows an arithmetic shift left and right by one bit position.

Figure 6-19. Arithmetic Shift Left and Right

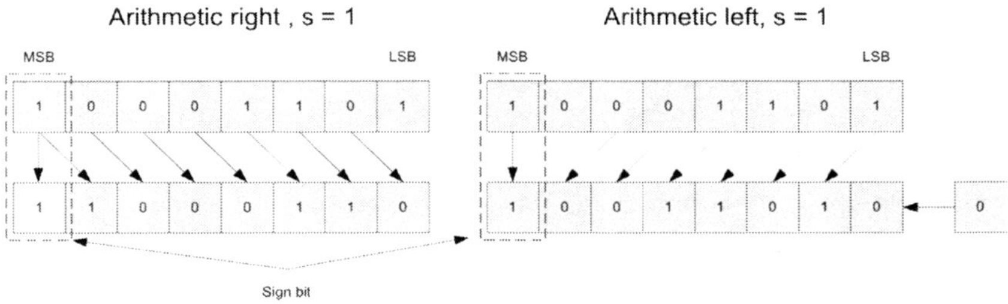

Example 6-32 show a C++ design that is inferred as an arithmetic right shift. It can be seen that the only difference from Example 6-30 is that the data variable and return type are both signed.

Example 6-32. Barrel Shifter with Arithmetic Shift Right

```
#include "barrel_shift_ar.h"
ac_int<NUM_BITS,true> barrel_shift_ar(ac_int<NUM_BITS,true>
din,
                                                     ac_int<CTRL_
BITS,false> s){
  return din >> s;
}
```

Bi-directional

Creating a bi-directional barrel shift for either arithmetic or logical shifting is as easy as changing the shift variable "s" in the previous examples to signed. When this is done positive values of "s" shift in the direction specified in the C++, negative values shift in the opposite direction. The number of bits for "s" must be increased by one to account for the sign bit since "s" can now be negative. This means that "s" requires log2ceil(NUM_BITS) + 2 bits. Example 6-33 shows the implementation of the bi-directional arithmetic barrel shifter. Positive values of "s" shift right, negative values of "s" shift left.

Example 6-33. Bi-directional Barrel Shifter

```
#include "barrel_shift_bidir_a.h"
ac_int<NUM_BITS,true> barrel_shift_bidir_a(ac_int<NUM_BITS,true>
din, ac_int<CTRL_BITS,true> s){
    return din>>s;
}
```

Rotating

Certain applications such as encryption require a barrel shifter that rotates the data, preserving the bits that are normally discarded from either the MSB or LSB position. This can easily be realized by combining a logical left and right barrel shifter as shown in Example 6-34.

Example 6-34. Rotating Barrel Shifter

```
1   #include "rotate_r.h"
2   ac_int<NUM_BITS,false> rotate_r(ac_int<NUM_BITS,false> din,
3                                   ac_int<CTRL_BITS,false> s){
4     return (din >> (s%NUM_BITS)) | (din << (NUM_BITS-(s%NUM_
BITS)));
5   }
6
```

The rotating barrel shifter design is designed for any size bit-vector, and it has built-in protection to guarantee that the rotate is always done correctly for non power-of-two bit vectors. This is accomplished by using the modulus operator "%" to make sure that neither shift value exceeds the maximum number of bits "NUM_BITS". However non power-of-two bit vectors cause additional adder logic to be inferred for the "%" operator. The "%" can be removed if "s" is never allowed to range beyond NUM_BITS. If the protection of the modulus operator is required there is a slightly better way to code Example 6-34 for improved area. Line 4 of Example 6-34 contains two instances of the expression "s%NUM_BITS". These two expressions are optimized first using sequential constant propagation because NUM_BITS is a constant. Unfortunately this prevents efficient sharing of common sub-expressions. Example 6-35 shows a better way to code the rotating barrel shifter so that the sub-expression "s%NUM_BITS" is shared. A temporary variable "stmp" is used on line 4 to compute "s%NUM_BITS". This variable is then used twice on line 5, explicitly forcing the sharing of the sub-expression.

Example 6-35. Improved Rotating Barrel Shifter

```
#include "rotate_r.h"
ac_int<NUM_BITS,false> rotate_r(ac_int<NUM_BITS,false> din,
                                ac_int<CTRL_BITS,false> s){
  ac_int<CTRL_BITS,false> stmp = s%NUM_BITS;
  return (din >> stmp) | (din << (NUM_BITS-stmp));
}
```

Constant Shifts

The previous section illustrated that a barrel shifter is inferred when the shift control is programmable. This means that the barrel shift hardware is constructed so that the input bit vector can be arbitrarily shifted from 0 to 2^CTRL_BITS bits. This may not always be necessary if only a subset of shift values are required. When this is the case, it is better to re-write the design to use constant shifts. The improvement on area depends on not only the size of the bit vector, but the required number of shifts.

Transforming Barrel Shifters into Constant Shifts

Consider the "Barrel Shifter with Logical Shift Right" on page 146 with it constrained such the shift control "s" can only take on one of three values, "0", "1", and "5". The design can be re-coded to improve area by taking advantage of the fact that there are three constant shifts, shown in Example 6-36.

Example 6-36. Transforming Barrel Shifters into Constant Shifts

```
#include "barrel_shift_lr.h"
ac_int<NUM_BITS,false> barrel_shift_lr(ac_int<NUM_BITS,false> din,
                                       ac_int<CTRL_BITS,false> s){
  ac_int<NUM_BITS,false> tmp = din;

  if(s==1)
    tmp >>= 1;
  else if(s==5)
    tmp >>= 5;

  return tmp;
}

Design Constraints
NUM_BITS = 8, CTRL_BITS = 4
```

The details of Example 6-36 are:

- Line 4 uses a temporary variable to read "din".
- Lines 6 through 9 test the shift variable "s" against the allowed bits shifts, and if true shifts "tmp" by a constant value. If nothing matches then "tmp" is returned unshifted which is the same as shift by zero.

Although Example 6-36 is an improvement in area over Example 6-30, it still has some inefficiencies. The shift control variable "s" is "CTRL_BITS", 4 bits in this example, wide even though there are only three possible shift values. Instead of using the explicit shift value, "s" can be encoded to select one of three shift possibilities. This should reduce the amount of comparison logic and can make a substantial impact on large shifters. Example 6-37 illustrates this technique. In this case two bits can encode the three possible shift value.

Example 6-37. Encoding the Shift Control

```
#include "barrel_shift_lr.h"
ac_int<NUM_BITS,false> barrel_shift_lr(ac_int<NUM_BITS,false> din,
                                                   ac_int<2,false> s){
  ac_int<NUM_BITS,false> tmp = din;

  if(s==0)
    tmp >>= 1;
  else if(s==1)
    tmp >>= 5;

  return tmp;
}
```

Transforming Dynamic Bit Masking

Another cause for the unwanted inferencing of barrel shifters is when the iterator of an rolled loop is used to shift and mask a bit vector. This is often done when trying to count the number of ones in a bit vector, or when performing distributed arithmetic type operations. Example 6-38 shows a design that uses dynamic shifting to count the ones in a bit vector. The expression "din>>i" on line 5 causes a barrel shifter to be inferred since the loop is left rolled.

Example 6-38. Dynamic Bit Masking

```
#include "shift_mask.h"
ac_int<RES_BITS,false> test(ac_int<NUM_BITS> din){
  ac_int<RES_BITS,false> acc=0;
  for(int i=0;i!=NUM_BITS;i++)
    acc += (din>>i)&1;
  return acc;
}
```

```
Design Constraints
Main loop pipelined with II=1
All loops left rolled
```

Example 6-38 can be rewritten to eliminate the need for a barrel shifter by storing the input and shifting it by one bit during each iteration of the loop, shown in Example

Example 6-39. Static Bit Masking

```
1   #include "shift_mask.h"
2   ac_int<RES_BITS,false> test(ac_int<NUM_BITS> din){
3     ac_int<RES_BITS,false> acc=0;
4     ac_int<NUM_BITS>tmp = din;
5
6     for(int i=0;i!=NUM_BITS;i++){
7       acc += tmp&1;
8       tmp >>= 1;
9     }
10    return acc;
11  }
```

The details of Example 6-39 are:

- Line 4 reads the input "din" and stores it in a temporary variable.
- Line 7 masks off the LSB of "tmp" and adds it to "acc".
- Line 8 shifts "tmp" by one bit to the right.

Adder Trees

Automatic Tree Balancing

High Level Synthesis always tries to build a balanced tree structure out of a number of related additions that can be scheduled in parallel. The most typical case of this is accumulating the elements of an array inside of a loop that is fully unrolled, or accumulating the products of the taps and coefficients of a FIR filter. Balancing the adder tree tends to help reduce the area of a design by minimizing the latency, which in turn reduces the number of registers. Example 6-40 shows a design that results in a balanced adder tree.

Example 6-40. Automatic Tree Balancing

```
#include "balanced.h"
ac_int<WIDTH_OUT,false> balanced(ac_int<WIDTH,false>
din[NUM_REGS]){
  ac_int<WIDTH_OUT,false> acc = 0;

  for(int i=0;i!=NUM_REGS;i++)
    acc += din[i];
  return acc;
}


Design Constraints
Main loop pipelined with II=1
All loops fully unrolled
WIDTH=8, NUM_REGS=8
```

Figure 6-20 shows the synthesized adder tree assuming that there is just enough time to schedule it within one clock cycle.

Figure 6-20. Balanced Adder Tree

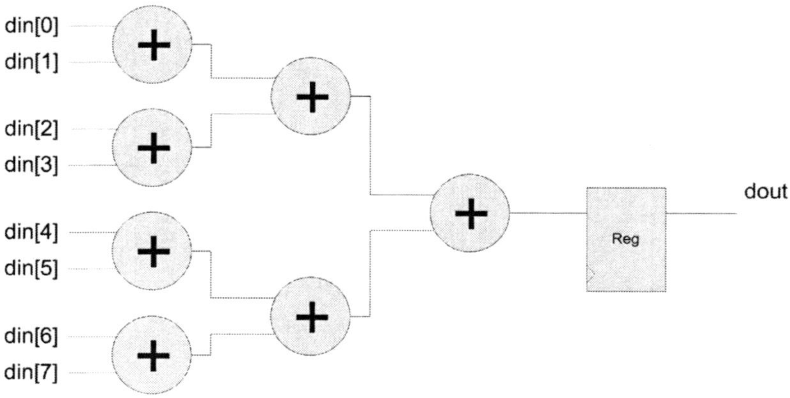

Preventing Automatic Tree Balancing

Automatic tree balancing sometimes is prevented when the accumulate inside the unrolled loop is controlled by a condition. Simple conditions that don't change between loop iterations can usually be balanced. However, when the condition changes between each iteration it is likely that the tree is not balanced. This can have a very negative impact on both area and performance. Example 6-41 shows just such a case where the elements of "din" are accumulated based on the element of "s" being set to true.

Example 6-41. Preventing Automatic Tree Balancing

```
1   #include "unbalanced_tree.h"
2   ac_int<WIDTH_OUT,false>  unbalanced(ac_int<WIDTH,false>
din[NUM_REGS],
3                                 bool s[NUM_REGS]){
4     ac_int<WIDTH_OUT,false> acc = 0;
5     ac_int<WIDTH,false> tmp[NUM_REGS];
6
7     for(int i=0;i!=NUM_REGS;i++)
8       tmp[i] = din[i];
9     for(int i=0;i!=NUM_REGS;i++)
10      if(s[i])
11        acc += tmp[i];
12
13    return acc;
14  }

Design Constraints
Main loop pipelined with II=1
All loops fully unrolled
WIDTH=8, NUM_REGS=8
```

The details of Example 6-41 are:

- Lines 7 and 8 - "din" is copied unconditionally into temporary storage "tmp". This was done because the example design is synthesized as the top-level design. Having conditional IO in a pipelined design often causes scheduling to fail. Making an internal copy of din would not be necessary if this function was being called somewhere below the top-level.
- Lines 9 through 11 - each "din[i]" is accumulated when "s[i]" is true. Tree balancing is prevented since each loop iteration depends on a different "s[i].

The resulting hardware from Example 6-41 is shown in Figure 6-21. In this case an adder chain with "NUM_REGS" levels of logic is scheduled as opposed to log2ceil(NUM_REGS) levels of logic. This may lead to longer latency and larger area for higher clock frequencies.

Figure 6-21. Unbalanced Adder Chain

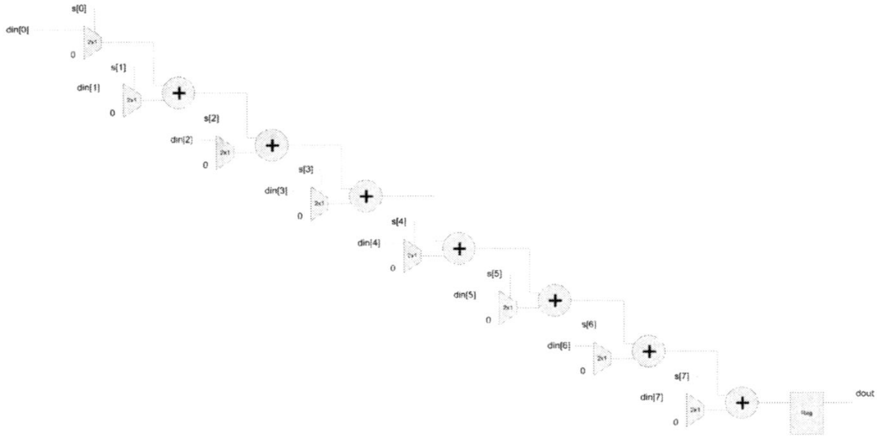

Coding to Facilitate Automatic Tree Balancing

The best way to facilitate adder tree balancing is to make the adds unconditional. The question mark operator "?" is usually used to accomplish this as shown in Example 6-42. In this example, on Line 10, the accumulate is performed for every iteration of the loop regardless of the value of "s[i]". "s[i]" is used as the selection variable for the question mark operator to add either "tmp[i]" or zero. In other words the add has been made unconditional. The resulting hardware is shown in Figure 6-22.

Example 6-42. Forcing Adder Tree Balancing

```
1   #include "rebalanced.h"
2   ac_int<WIDTH_OUT,false>  rebalanced(ac_int<WIDTH,false>
din[NUM_REGS],
3                                bool s[NUM_REGS]){
4     ac_int<WIDTH_OUT,false> acc = 0;
5     ac_int<WIDTH,false> tmp[NUM_REGS];
6
7     for(int i=0;i!=NUM_REGS;i++)
8       tmp[i] = din[i];
9     for(int i=0;i!=NUM_REGS;i++)
10      acc += s[i] ? tmp[i] : 0;
11
12    return acc;
13  }

Design Constraints
Main loop pipelined with II=1
All loops fully unrolled
WIDTH=8, NUM_REGS=8
```

Figure 6-22. Forcing Adder Tree Balancing

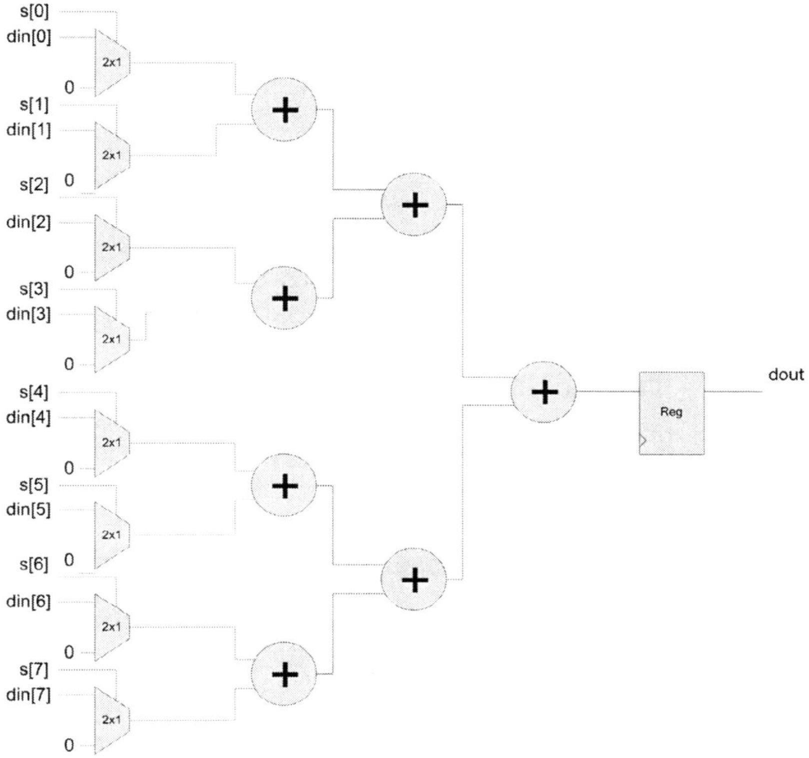

Lookup Tables (LUT)

HLS infers lookup tables from constant arrays. The hardware realized for the LUT can be either a MUX with constant inputs, or a ROM. The choice of one over the other is user defined, with the default behavior being a MUX based implementation. C++ makes the parameterization, generation and inclusion of lookup tables very simple. Example 6-43 shows a LUT based implementation of sin(x).

Example 6-43. Lookup Table for sin(x)

```
#include "lut.h"
ac_fixed<WIDTH,2> lut(ac_int<ADDR_WIDTH,false> i){
  const ac_fixed<WIDTH,2> sin_table[NUM_REGS] = {
    #include "data.inc"
  };
  return sin_table[i];
}
```

The details of Example 6-43 are:

- Line 3 declares a constant array of type ac_fixed<WIDTH,2>. Although it is not strictly necessary to declare the array as "const" to infer a lookup table, it is considered good programming style, and may help in standard C++ compilation. The reason why the data is declared with 2 integer bits, and the rest fractional, is to represent an sin(x) value between -1 and 1.
- Line 4 uses a convenient technique in C++ that allows the inclusion of a text file "data.inc" that specifies all of the constant values for sin_table by having the #include is within the braces "{ }" for "sin_table"

Example 6-44 shows the header file "lut.h" included in Example 6-43. The header file allows parameterization of the number of lookup table elements and uses the lod2ceil helper class to compute the required number of address bits.

Example 6-44. Lookup Table Header File

```
#ifndef __LUT__
#define __LUT__
#include <ac_fixed.h>
#include "../../helper_classes/src/log2ceil.h"

#define WIDTH 8
#define NUM_REGS 16
#define ADDR_WIDTH LOG2_CEIL<NUM_REGS>::val
ac_fixed<WIDTH,2> lut(ac_int<ADDR_WIDTH,false> i);

#endif

```

The actual "sin_table" constants for Example 6-43 are generated using a separate C++ program, shown in Example 6-45.

Example 6-45. Lookup Table Generation

```
#include <ac_fixed.h>
#include <math.h>
#include <iostream.h>
#include <fstream.h>
#include "lut.h"

int main(){
  ac_fixed<WIDTH,2> data;
```

```
9      double pi = 3.14159265358979323846264338327950;
10     fstream fptr;
11
12     fptr.open("data.inc",fstream::out);
13
14     for(int i=0;i<NUM_REGS;i++){
15       data = sin(2*pi*i/(double)NUM_REGS);
16       fptr << data;
17       if(i != NUM_REGS-1)
18         fptr << ", " << endl;
19     }
20
21     fptr.close();
22
23   }
```

The details of Example 6-45 are:

- Line 5 includes the "lut.h" header file. This is where the #defines for WIDTH and NUM_REGS are declared for the actual lookup table design. Doing this allows the table generation to be matched to the C++ implementation.
- Line 12 opens a text file "data.inc" for writing.
- Line 15 generates sin values from 0 to 2*pi
- Lines 17 and 18 insert commas between data values except for the last value

The generated table data is shown below in Example 6-46.

Example 6-46. Generated Lookup Table Data

```
0,
.375,
.703125,
.921875,
1,
.921875,
.703125,
.375,
0,
-.390625,
-.71875,
-.9375,
-1,
-.9375,
-.71875,
-.390625
```

Lastly the Makefile for this design can be written so that the table generation and design compilation are dependent on one another. This allows the design to be recompiled successfully when the design parameters are changed, shown in Example 6-47.

Example 6-47. Lookup Table Makefile with Dependencies

```
1  #MACROS
2  CAT_HOME   = $(MGC_HOME)
3  CXX        =  /usr/bin/g++
4  CXXFLAGS    =  -g  -O  -Wall  -Wno-deprecated   $(DEFINES) $(INCLUDES)
5  INCLUDES   = -I "$(CAT_HOME)/shared/include"
6
7  TARGET0     = gen_tbl
8  OBJECTS0    = gen_sin_table.o
9  DEPENDS0    = Makefile lut.h
10 $(TARGET0): $(OBJECTS0)
11     $(CXX) $(CXXFLAGS) -o $(TARGET0) $(OBJECTS0)
12 $(OBJECTS0): $(DEPENDS0)
13
14 TARGET1     = tb
15 OBJECTS1    = tb_lut.o lut.o
16 DEPENDS1    = Makefile lut.h data.inc
17 $(TARGET1): $(OBJECTS1)
18     $(CXX) $(CXXFLAGS) -o $(TARGET1) $(OBJECTS1)
19 $(OBJECTS1): $(DEPENDS1)
20
21 .PHONY: run
22 run0: $(TARGET0)
23     ./gen_tbl.exe
24 .PHONY: run1
25 run1: $(TARGET1)
26     ./tb.exe
27 #phony target to make and run table generation and design and tb
28 .PHONY: all
29 all: run0 run1
30
31 #phony target to remove all objects and executables
32 .PHONY: clean
33 clean:
34     rm -f *.o *.exe *.inc
35
```

The details of the Makefile in Example 6-47 are:

- Lines 9 and 16 make both the table generation and the test bench and design dependent on "lut.h". This forces recompilation of both designs if the parameters are changed.

- Line 16 is also dependent on the table data itself "data.inc" and is recompiled if a change is made.
- Lines 28 and 29 builds the target for the table generation, generates the table, and then builds the target for the test bench and design

References

1. Andres Takach, David Burnette, and Michael Fingeroff. C++ IP Design and Reuse. DesignCon 2009.

Chapter 7

Memory Architecture

Introduction

Up until now the previous chapters have focused primarily on algorithms that have register based memory architectures. This reasons for this were twofold; one being that the hardware building blocks covered in "Sequential and Combinational Hardware" on page 113 are typically implemented using registers. More importantly it allowed the introduction of HLS concepts such as loop unrolling and pipelining as a means for exploring parallelism without having to consider how array access patterns may prevent scheduling when arrays are mapped to memories instead of registers. Although it has be said before, it's worth repeating, that HLS often gives you exactly what you asked for. Arrays mapped to memories tend to be the bottleneck in a design's performance. HLS provides a number of automatic optimizations and constraints, such as memory splitting, interleaving, and merging, that can remove these memory bottlenecks. Whenever possible, these automatic memory optimizations should be used, minimizing the number of code modifications. However, there may be situations where explicitly coding the memory architecture is either required to meet performance, or may allow designers to achieve even better quality of results. In these cases it is essential that array accesses are coded in such a way as to not limit performance. This means analyzing array access patterns and organizing the memories in a design so that the desired throughput and area can be achieved.

Memory-based Shift Register

A shift register implemented using memories is a good starting point to understand the impact that array access patterns have on performance. Because it's memory based, the read of the shift register taps cannot occur in parallel, which means that loops must be left fully or partially rolled if used in the implementation. A memory based shift register would typically be used in something like a FIR filter with a very large number of taps, where it is impractical to use registers because of the area and power costs.

Classic Shift Register Description mapped to Memories

If we revisit the classic register-based shift register, it becomes obvious why its memory architecture is unsuitable for a memory-based implementation.

Example 7-1. Register-based Shift Register

```
#include "basic_shift.h"
void shift_reg(dType din, dType dout[N_REGS]){
  static dType regs[N_REGS];
 SHIFT:for(int i=N_REGS-1;i>=0;i--){
    if(i==0)
      regs[i] = din;
    else
      regs[i] = regs[i-1];
  }
 WRITE:for(int i=0;i<N_REGS;i++)
    dout[i] = regs[i];
}
```

```
Design Constraints
Main loop cannot be pipelined
"din" mapped to registers or wire interface
All loops left rolled
"regs" array mapped to RAM with separate read and write ports
N_REGS = 4
```

The schedule for the SHIFT loop in Example 7-1 is shown below in Figure 7-1.The rest of the design schedule has been excluded to simplify the scheduling diagram.

Figure 7-1. Schedule for Classic Shift Register Mapped to RAM

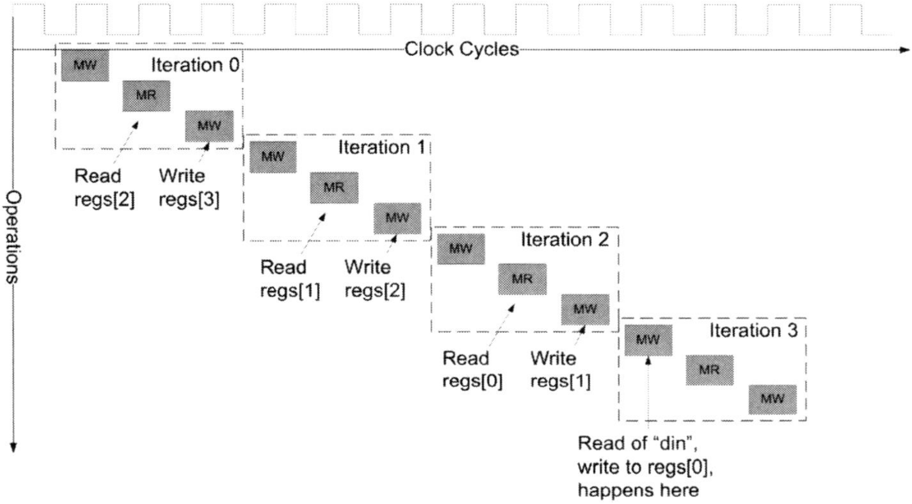

What Figure 7-1 illustrates is that the register-based description of a shift register is inefficient when arrays are mapped to memory. It takes about 12 clock cycles to shift all four taps. This is because each tap that is shifted requires reading then writing the memory. Furthermore, the conditional write of "regs[0]" on line 6 of Example 7-1 causes an additional write operation to be scheduled for each loop iteration, even though the condition only evaluates to true when i==0. Although this extra write could be eliminated by using a temporary variable to choose between writing "din" or regs[i-1], it would not solve the bigger problem that the read/write array access pattern is not efficient when the array is mapped to memory.

Note

C++ written for register based memory architectures is often unsuitable for memory based architectures.

Circular Buffer

The most efficient implementation for a memory-based shift register is to use the same approach that would be used when writing code for a micro-processor. In this case read and write pointers can be used to implement a circular buffer. This moves the write and read pointer locations as new data is shifted in, rather than moving the data for each tap. Figure 7-2 shows how the read and write pointers circulate. Note that the read pointer runs in the opposite direction as the write pointer.

Figure 7-2. Circular Buffer Pointer Motion

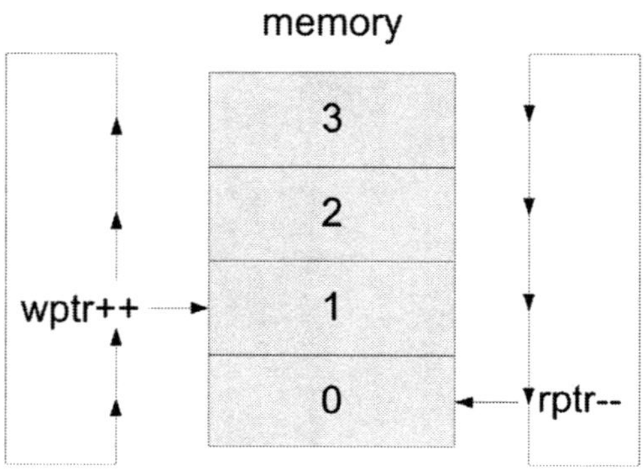

Example 7-2 shows a circular buffer implementation of a shift_register class.

Example 7-2. Circular Buffer Shift Register Class

```
#ifndef __SHIFT__
#define __SHIFT__
#include <ac_int.h>
template<typename T, int N>
  class circular_shift{
 private:
  T mem[N];
  ac_int<ac::log2_ceil<N>::val+1,false> wptr;
  ac_int<ac::log2_ceil<N>::val+1,true> rptr;
 public:
  circular_shift(){
    T dummy;
    wptr = 0;
    rptr = 0;
    for(int i=0;i<N;i++)
      mem[i] = dummy;
  }
  void operator <<(T data){
    mem[wptr] = data;
    wptr++;
    if(wptr==N)
      wptr=0;
  }
  T operator [](ac_int<ac::log2_ceil<N>::val,false> idx){
    rptr = (wptr-1-idx);
```

```
26      if(rptr<0)
27        rptr = rptr+N;
28      return mem[rptr];
29    }
30  };
31  #endif
```

The details of Example 7-2 are:

- Line 4 allows both the type and number of shift register elements to be specified as a template parameter.
- Lines 8, 9, and 24 uses the log2_ceil function that is built in to the ac_int bit accurate data types library to compute the minimum number of bits based on the array size “N”. In “Sequential and Combinational Hardware” on page 113, the log2ceil was computed using helper classes to illustrate the usefulness of classes and enumerated types for computing static values based on template parameters. Since log2ceil is already available in the ac_int.h library it is used from this point forward.
- Lines 11 through 17 - The class constructor is used to initialize the read and write pointers to zero and to initialize the “mem” array” to don’t care. See Initialization loops for more detail on un-initializing arrays mapped to memories.
- Lines 18 through 23 implement the shift operator “<<“. The writes to “mem” are limited to one write per call to “<<” since “mem” is mapped to memory. Each shift cause a new data value to be written into “mem[wptr]” after which “wptr” is incremented. When the end of the memory is reached, line 21, the write pointer “wptr” is moved back to the beginning of the memory. Hence the write pointer continuously circulates through the memory addresses.
- Lines 24 through 30 implement the “[]” operator for reading the shift register taps. The read pointer “rptr” runs in the opposite direction as the write pointer. If the read pointer becomes negative, line 26, it is moved to the top of the memory.

Initialization loops

Note

Care should be taken when mapping arrays to memory if the array has been declared static. C++ requires that static arrays or variables are reset to zero, which in turn causes the creation of an "initialization loop" when a design is synthesized with static arrays mapped to memory.

This in turn implies that hardware is generated to step through every address of the memory and set the data stored at that location equal to zero. Not only does this cost extra cycles of latency when the design is reset, but also increases area, and may limit pipelining. In many cases a memory does not have to be reset to zero because it is known that it is written before it is read. For situations such as these it is desirable to remove the initialization loop, while still leaving the array declared static. To do this the array must be initialized to "don't care". Lines 11 through 17 of Example 7-2 shows how this is done in the class constructor. Line 12 defines a variable "dummy" which is left un-initialized (don't care). This variable is then assigned to all elements of "mem" on lines 15 and 16. Doing this removes the initialization loop from the design. Example 7-2 explicitly codes the un-initialization if the array into the constructor based on the data type "T". There are built-in utility functions in the Algorithmic C libraries that can be used if the data type is either native C++ or ac_int or ac_fixed (See "Helper/Utility Functions" on page 33).

Memory Organization

The introduction to this chapter discussed the abilities of HLS to automatically solve memory bottleneck problems by allows memories to be split, interleaved, or reorganized. However, there often situations where these automatic optimizations may not be optimal. This is often caused by having arrays, or operations on arrays, that are not a power-of-two. This section deals with how the C++ code, and memory access, should be manually reorganized if the automatic optimizations do not provide adequate results.

Interleaving Memories

Interleaving in hardware design is the process of rearranging sequential data storage into two or more non-contiguous storage blocks to increase performance.

Automatic Interleaving

The reasons for interleaving memory accesses becomes apparent by examining Example 7-3.

Example 7-3. Accessing Multiple Array/Memory Locations

```
1  #include "interleave.h"
2  void interleave(ac_int<8> x_in[NUM_WORDS],    ac_int<8>
y[NUM_WORDS/3],
3                  bool load){
4    static ac_int<8> x[NUM_WORDS];
5    int idx = 0;
6
7    if(load)
8      for(int i=0;i<NUM_WORDS;i+=1)
9        x[i] = x_in[i];
10   else
11     for(int i=0;i<NUM_WORDS;i+=3)
12       y[idx++] = x[i]+x[i+1]+x[i+2];
13
14 }

Design Constraints
"x_in", x, and y mapped to singleport memory
All loops left rolled
NUM_WORDS = 9
```

The details of Example 7-3 are:

- Line 4 defines an internal array "x" which is mapped to memory.
- Line 7 tests "load", and if true loads the internal memory.
- Lines 11 and 12 increment the index in multiples of three and adds three sequential values of "x". Using the specified constraints, this design cannon be pipelined with II=1 because "x" is mapped to a singleport RAM. Three separate reads from "x" are required each clock cycle, which is not possible from a singleport RAM.

If automatic interleaving is set on "x", it can be partitioned into three separate singleport RAMs as shown in Figure 7-3. The memories are organized so that each one of the three reads on Line 12 of Example 7-3 occur from a separate memory, allowing the design to run with II=1. Although the use of automatic memory interleaving may be sufficient, manually coding it into the design usually results in smaller area when the interleaving factor is not a power of two. In this design example the interleave factor is by three.

Figure 7-3. Interleaving by Three

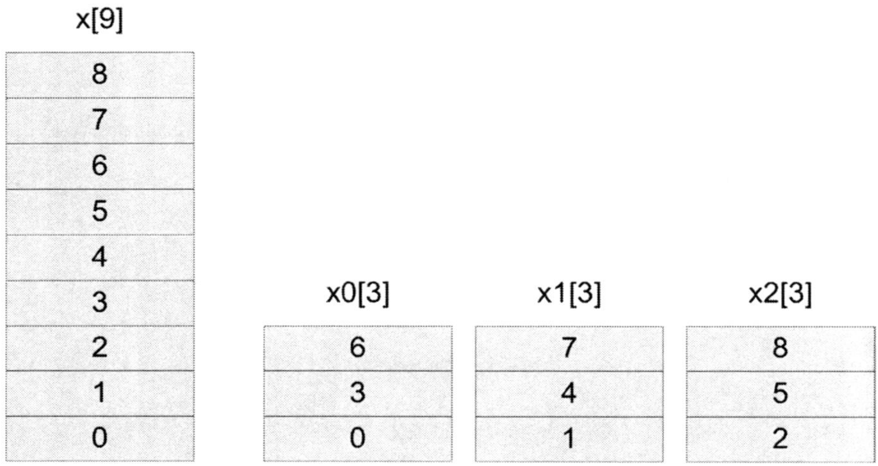

Manual Interleaving with Random Access

Example 7-3 can be rewritten with relative ease to manually interleave three memories, while still maintaining random access into the original array “x”. Doing this has the advantage of reducing area by explicitly coding the interleaving control into the C++, which can yield better results than automatic interleaving for non power of two interleaving. A class based approach can be taken to encapsulate the interleaved memory architecture, helping to minimize the code changes in the original algorithm. Example 7-4 shows the rewritten C++.

Example 7-4. Manual Interleaving with Random Access

```
#include "interleave.h"
#include "interleave_mem.hpp"
void interleave_manual(ac_int<8> x_in[NUM_WORDS],
                       ac_int<8> y[NUM_WORDS/3], bool load){
  static interleave_mem<ac_int<8>,NUM_WORDS> x;
  int idx = 0;

  if(load)
    for(int i=0;i<NUM_WORDS;i+=1)
      x.write(i,x_in);
  else
    for(int i=0;i<NUM_WORDS;i+=3)
      y[idx++] = x.read(i,0) + x.read(i,1) + x.read(i,2);
}
```

```
Design Constraints
"x_in", x, and y mapped to singleport memory
All loops left rolled
All sub-loops pipelined with II=1
NUM_WORDS = 9
```

The details of Example 7-4 are:

- Line 5 declares a static instance of a class that implements memory interleaving by three. The class takes both the data type and number of array elements as its template arguments.
- Line 10 calls the interleave memory class "write" method and passes the index value "i" and the input array "x_in".
- Line 13 accesses the data from the interleaved memory class using the "read" method. The "read" method takes the index "i" and the constant offset as separate function arguments. Passing the constant offset as a separate argument allows simultaneous scheduling of mutually exclusive memory reads. This is discussed further when delving into the details of the interleaved memory class.

Example 7-5 shows the class definition for the interleaved memory class.

Example 7-5. Interleaved Memory Class with Random Access

```
#ifndef __INTERLEAVE_MEM__
#define __INTERLEAVE_MEM__
#include <ac_int.h>
template<typename T, int N>
  class interleave_mem{
  T x0[N/3];
  T x1[N/3];
  T x2[N/3];
 public:
  interleave_mem(){
  }
   void write(ac_int<ac::log2_ceil<N>::val,false> i, T x_in[N]);
  T read(ac_int<ac::log2_ceil<N>::val,false> i, int offset);
};
#include "read_mem.hpp"
#include "write_mem.hpp"
#endif
```

The details of Example 7-5 are:

- Line 4 allows the type and number of array elements to be specified as a class template parameter.
- Lines 6 through 8 define three separate arrays with N/3 elements. Note that there is no check here to guarantee that N is evenly divisible by three, so the designer would have to make sure that this is instantiated correctly. Alternatively the class could be enhanced to support any value for N.
- Lines 12 and 13 define the class read and write methods and use the log2_ceil helper functions from the ac_int data type library to make sure that the index "i" is reduced to the minimum number of bits.
- Lines 15 and 16 include the header files that implement the class read and write methods. Usually the code for these methods would be inlined in the same header file, but they are kept separate in this style guide so that the size of any one piece of code under discussion is kept as small as possible.

Example 7-6 shows the implementation of the "write" method for the interleaved memory class.

Example 7-6. Interleaved Memory Class Random Access Write Method

```
#ifndef __WRITE_MEM__
#define __WRITE_MEM__
#include <ac_int.h>
template<typename T, int N>
void interleave_mem<T,N>::write(ac_int
<ac::log2_ceil<N>::val,false> i, T x_in[N]){
    T tmp = x_in[i];
    switch(i%3){
    case 0:
      x0[i/3] = tmp;
      break;
    case 1:
      x1[i/3] = tmp;
      break;
    case 2:
      x2[i/3] = tmp;
      break;
    }
  }
#endif
```

The details of Example 7-6 are:

- Line 7 reads "x[i]" into a temporary variable to limit reading of the array mapped to memory to once per clock cycle.
- Line 8 selects between one of the three memories by taking the mod3 of the index "i".
- Lines 9 through 18 implement the writing of the three internal memories, x0, x1, and x2. The original index "i" is divided by three since each memory has N/3 elements. The use of both the constant divide and the constant modulus can be costly in terms of bigger area. It is often possible to eliminate these if the access to the memory is known to always be sequential rather than random access. This is discussed in the next section.

Example 7-7 shows the implementation of the "read" method for the interleaved memory class.

Example 7-7. Interleaved Memory Class Random Access Read Method

```
1  #ifndef __READ_MEM__
2  #define __READ_MEM__
3  #include <ac_int.h>
4  template<typename T, int N>
5  T   interleave_mem<T,N>::read(ac_int
6  <ac::log2_ceil<N>::val,false> i, int offset){
7      T tmp=0;
8      switch(offset){
9      case 0:
10       tmp = x0[i/3];
11       break;
12     case 1:
13       tmp = x1[i/3];
14       break;
15     case 2:
16       tmp = x2[i/3];
17       break;
18     }
19     return tmp;
20   }
21 #endif
```

The details of Example 7-7 are:

- Line 8 uses the "offset" input argument to select between the three arrays defined in the class. Constant propagation guarantees that only one memory read is scheduled per call to "read" since the "offset" is always passed as a constant (Shown in Example7-4).

- Lines 9 through 18 implement the three separate memory reads based on "offset". The index "i" is divided by three for each of the internal arrays (x0,x1,x2) since each array has N/3 elements and the incoming index ranges from 0 to N. This constant divide can be costly in terms of area. In many cases it can be eliminated if the design always accesses the memory in sequential order. This is covered in the next section.

Manual Interleaving with Sequential Access

The previous section "Manual Interleaving with Random Access" on page 165 showed how arrays mapped to memories can be manually interleaved to give the best possible performance as well as improved area over automatic interleaving. However, the algorithm that used the interleaved memory class in Example 7-4 on page 165 did not require random access into the array, which means that the interleaved memory class was overbuilt for the application. The interleaved memory class in Example 7-4 always writes and reads in sequential order. This behavior makes it possible to eliminate the constant modulus and constant divide operations used in the "write" and "read" methods of the class. Example shows Example 7-4 rewritten to exploit the sequential nature of the memory accesses.

Example 7-8. Manual Interleaving with Sequential Access

```
1   #include “interleave.h”
2   #include “interleave_mem_improved.hpp”
3   void interleave_manual(ac_int<8> x_in[NUM_WORDS],
4                          ac_int<8> y[NUM_WORDS/3], bool load){
5     static interleave_mem<ac_int<8>,NUM_WORDS> x;
6     int idx = 0;
7
8     if(load)
9       for(int i=0;i<NUM_WORDS;i+=1)
10        x.write(i,x_in);
11    else
12      for(int i=0;i<NUM_WORDS/3;i+=1)
13        y[idx++] = x.read(i,0) + x.read(i,1) + x.read(i,2);
14  }
```

The only change that was made to the algorithm itself is on line 12 of Example 7-8. In the original algorithm the loop iterated from 0 to NUM_WORDS incrementing by three. Instead the loop can be incremented from 0 to NUM_WORDS/3 by one since each call to "x.read" accesses one of three arrays with NUM_WORDS/3 elements. This allows the removal of the constant divider in the "read" method".

Example 7-9 shows the interleaved memory class definition rewritten to take advantage of the sequential nature of the array accesses.

Example 7-9. Interleaved Memory Class with Sequential Access

```
1   #ifndef __INTERLEAVE_MEM__
2   #define __INTERLEAVE_MEM__
3   #include "interleave.h"
4   template<typename T, int N>
5   class interleave_mem{
6     int x0[N/3];
7     int x1[N/3];
8     int x2[N/3];
9     ac_int<ac::log2_ceil<N>::val,false> idx;
10    ac_int<2,false> sel;
11  public:
12    interleave_mem(){
13      idx=0;
14      sel = 0;
15    }
16     void  write(ac_int<ac::log2_ceil<N>::val,false>  i,  T
x_in[N]);
17    T read(ac_int<ac::log2_ceil<N>::val,false> i, int offset);
18  };
19  #include "write_mem_improved.hpp"
20  #include "read_mem_improved.hpp"
21  #endif
```

The details of Example 7-9 are:

- Line 9 defines an index that is used for the "read" method. The bit width of "idx" is set to the optimal number of bits, log2_ceil(N).
- Line 10 defines a two bit counter variable "sel" that is used to select the current memory for writing.
- Lines 12 through 15 define the default constructor to reset the counters. This is required and is based on the assumption that the sequential reads and writes begin at address location zero.

Example 7-10 shows the "write" method rewritten to exploit the sequential nature of the memory accesses.

Example 7-10. Interleaved Memory Class Sequential Access Write Method

```
1   #ifndef __WRITE_MEM__
2   #define __WRITE_MEM__
3   #include <ac_int.h>
4   template<typename T, int N>
5   void interleave_mem<T,N>::write(ac_int
6   <ac::log2_ceil<N>::val,false> i, T x_in[N]){
7     int tmp = x_in[i];
8       switch(sel++){
9       case 0:
10        x0[idx] = tmp;
11        break;
12      case 1:
13        x1[idx] = tmp;
14        break;
15      case 2:
16        x2[idx++] = tmp;
17        break;
18      }
19      if(idx==N/3)
20        idx = 0;
21      if(sel==3)
22        sel = 0;
23    }
24  #endif
```

The details of Example 7-10 are:

- Line 8 has been changed to select the array to be written based on the "sel" counter. Each time the "write method is called the case based on the value of "sel" is executed and "sel" is post-incremented. This has the exact behavior as "switch(i%3)" as long as the array is written sequentially starting from address zero.
- Lines 10, 13, and 16 writes the current using the "idx" counter. Once the third array, x2, is written "idx" is incremented. Using "idx" as the array index eliminates the need for the constant divider "i/3" and reduces area.
- Lines 19 and 20 reset "idx" back to zero when the last element of all the memories is written. Thus this memory architecture is customized to match the algorithm behavior and expects all memory addresses to be written starting from zero. If this were not true, additional control would be needed in the class to account for this.
- Lines 21 and 22 check "sel" and reset it back to zero once "x2" has been written. This also relies on sequential writes starting from address zero.

Example 7-11 shows the "write" method rewritten to account for sequential array accesses.

Example 7-11. Interleaved Memory Class Sequential Access Read Method

```
#ifndef __READ_MEM__
#define __READ_MEM__
#include <ac_int.h>
template<typename T, int N>
T interleave_mem<T,N>::read(ac_int
<ac::log2_ceil<N>::val,false> i, int offset){
  T tmp=0;
  switch(offset){
  case 0:
    tmp = x0[i];
    break;
  case 1:
    tmp = x1[i];
    break;
  case 2:
    tmp = x2[i];
    break;
  }
  return tmp;
}
#endif
```

The only change that was made to Example 7-11 was to remove the constant divider "i/3" for the array index. This was possible because Line 12 of Example 7-8 was changed to iterate to NUM_WORDS/3 incrementing by one instead of three.

Widening the Word Width of Memories

Automatic Word Width

Similar to interleaving some HLS tools allow designers to automatically widen the width of a memory so that data can be organized side by side. However, non-power of two array sizes, as well as read-modify-write issues sometimes make it desirable to manually code the word width expansion into the C++.

Manually Increasing Word Width with Sequential Access

Instead of interleaving memories, another possible solution to reading multiple sequential locations (Example 7-3) from a singleport memory every clock cycle is to widen the word width of the array mapped to memory. Then the sequential data can be written and read side-by-side. There are a few limitations to this approach. One is that there is a limit on how wide one can make the memory. And two is that writing a single value into the memory requires a read-modify-write, which may limit pipelining the design, especially if using true singleport RAM. Figure 7-4 shows how the data can be rearranged within the memory to place three words side by side. The memory is now three times as wide, but one third the number of elements.

Figure 7-4. Placing Words Side by Side in Memory

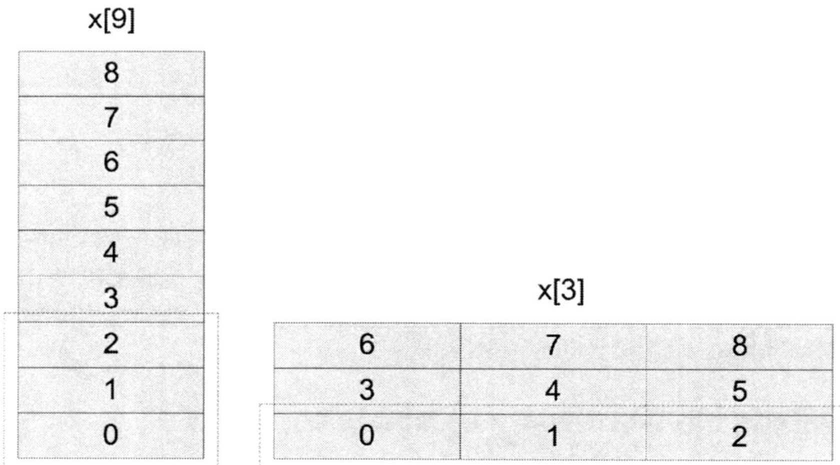

The original example used for demonstrating interleaving, "Accessing Multiple Array/Memory Locations" on page 164, is rewritten to show how to manually widen the word width. In this case it is assumed that the entire array is accesses sequentially from location zero to the end of the array. Example 7-12 shows the rewritten design using a memory class that widens the word width.

Example 7-12. Manually Widening the Word Width

```
#include "word_width.h"
#include "word_width_mem.hpp"
```

```
3   void word_width_manual(ac_int<8> x_in[NUM_WORDS],
4                          ac_int<8> y[NUM_WORDS/3], bool load){
5     static word_width_mem<8,true,NUM_WORDS> x;
6     int idx = 0;
7
8     if(load)
9       for(int i=0;i<NUM_WORDS;i+=1)
10        x.write(i,x_in);
11    else
12      for(int i=0;i<NUM_WORDS/3;i+=1)
13        y[idx++] = x.read(i,0) + x.read(i,1) + x.read(i,2);
14  }
```

The details of Example 7-12 are:

- Line 5 declares an instance of the "word_width_mem" class. This class takes the width, signedness, and number of array elements as the template arguments.
- Line 10 uses the class "write" method to write each element of the array.
- Line 12 adjusts the number of loop iterations to NUM_WORDS/3 and increments by one.
- Line 13 call the class "read" method and passes the index and the offset rather than "i+1", i+2". Similar to the interleaving example the word width class can take advantage of the offset specified as a constant to build more efficient hardware.

Example 7-13 shows the class definition for the "word_width_mem" memory class.

Example 7-13. Word Width Expansion Memory Class with Sequential Access

```
#ifndef __INTERLEAVE_MEM__
#define __INTERLEAVE_MEM__
#include <ac_int.h>
template<int W, bool S, int N>
class word_width_mem{
  ac_int<W*3,false> x[N/3];
  ac_int<ac::log2_ceil<N>::val,false> idx;
  ac_int<2,false> sel_rd;
  ac_int<2,false> sel_wr;
  ac_int<W*3,false> write3;
  ac_int<W*3,false> read3;
public:
  word_width_mem():sel_rd(0),sel_wr(0){}
  void write(ac_int<ac::log2_ceil<N>::val,false> i,
        ac_int<W,S> x_in[N]);
  ac_int<W,S> read(ac_int<ac::log2_ceil<N>::val,false> i,
```

```
17             const int offset);
18 };
19 #include "read_mem.hpp"
20 #include "write_mem.hpp"
21 #endif
```

The details of Example 7-13 are:

- Line 4 takes the width "W", the signedness "S", and the number of array elements "N" as the class template parameters. This class explicitly uses the ac_int data types in order to easily perform word width expansion.
- Line 6 defines an array "x" of unsigned ac_int that is W*3 times as wide as the original data and that has one third (N/3) as many elements. It should be noted that this class assumes that N is evenly divisible by three. If this was not true there would need to be more complicated control built into the class. There are no checks to test if this is true.
- Lines 8 and 9 define counters that are used for read and write slicing.
- Lines 10 and 11 define internal variables that are W*3 times as wide as the original data. These variables are used to access the array "x".
- Line 13 initializes the slice counters to zero.
- Lines 14 through 18 define the prototypes for the read and write methods.
- Lines 19 and 20 include the header files that define the read and write methods.

Example 7-14 shows the implementation of the word_width_mem class read method.

Example 7-14. Word Width Expansion Class Read Method

```
#ifndef __READ_MEM__
#define __READ_MEM__
#include <ac_int.h>
template<int W, bool S, int N>
ac_int<W,S> word_width_mem<W,S,N>::read
(
 ac_int<ac::log2_ceil<N>::val,false> i,
 const int offset
 ){
  ac_int<W,S> tmp=0;
```

```
11
12    if(sel_rd++==0)//read once every 3 calls
13      read3 = x[i];
14    if(sel_rd==3)
15      sel_rd = 0;
16    switch(offset){
17    case 0:
18      tmp = read3.template slc<W>(0);
19      break;
20    case 1:
21      tmp = read3.template slc<W>(W);
22      break;
23    case 2:
24      tmp = read3.template slc<W>(2*W);
25      break;
26    }
27    return tmp;
28  }
29  #endif
```

The details of Example 7-14 are:

- Line 5 - the read method returns an ac_int with the same width and signedness as the original data type.
- Line 7 uses log2_ceil to ensure that the minimum number of bits are used for the index "i".
- Lines 12 and 13 post increments "the read counter "sel_rd" and reads from the data member array "x" once every three calls to the read function. The data read from "x[i]" is stored in read3 which is W*3 bits wide. The index "i" has already been adjusted from the calling function to account for three words stored side-by-side.
- Line 14 checks the read counter to see when the read function has been called three times, and then resets it back to zero, initiating the next read of the array "x".
- Lines 16 through 26 checks "offset" passed to the read method and selects one of the three side-by-side data values. Note that each time the read method is called the sel_rd counter is advanced so this class expects all three offsets to be read. Otherwise the functionality does not match the original algorithm.

Example 7-15 shows the implementation of the word_width_mem class write method.

Example 7-15. Word Width Expansion Class Write Method

```
1   #ifndef __WRITE_MEM__
2   #define __WRITE_MEM__
3   #include <ac_int.h>
4   template<int W, bool S, int N>
5   void word_width_mem<W,S,N>::write
6   (ac_int<ac::log2_ceil<N>::val,false> i,
7    ac_int<W,S> x_in[N]
8    ){
9     write3.set_slc(sel_wr*W,x_in[i]);
10    sel_wr++;
11    if(sel_wr==3){
12      x[idx++] = write3;
13      sel_wr = 0;
14    }
15    if(idx==N/3)
16      idx = 0;
17  }
18  #endif
```

The details of Example 7-15 are:

- Line 9 - a write slice is made into “write3” each time the write method is called. “write3” is W*3 bits wide and is used as intermediate storage to store three sequential array elements side by side. This architecture assumes that all array elements are written in order for the entire array. Random access is not possible with this implementation and would require more complicated control. This memory architecture has been matched to the original algorithm.
- Lines 10 through 14 - the write counter “sel_wr” is incremented each time the write method is called. The data member array “x” is written when three array elements have been stored side-by-side into the intermediate storage “write3”. The write index “idx” is incremented on each write to “x”. Once the data is written the write counter is cleared and the next write begins.
- Lines 15 and 16 check to see when the write index reaches the end of the array/memory “x” and clears the index. Once again the assumption has been made that all array elements are written in sequential order, and the original array size “N” is evenly divisible by three. If this was not the case, and random access was required, more complicated control would need to be coded into the implementation.

Caching

Using True Single Port RAM as a Dualport RAM

A recurring theme when using arrays mapped to memories in high level C++ synthesis is that multiple simultaneous array accesses are often the cause of scheduling failures when trying to pipeline with II=1. This is even true when trying to implement something as simple as a line buffer using only singleport memories. A line buffer is used to store a single line of video from an image. The buffer must be able to be read and written every clock cycle. The simplest solution would be to map the line buffer storage array to a RAM that has both a read port and a write port. These types of RAM are usually referred to as RAM with separate read/write ports. The drawback to using these types of RAMs is that they require as much as 50% more area than a true singleport RAM. The problem with using singleport RAM is that it cannot be read and written in the same clock cycle, which makes implementing something like a line buffer a little tricky. A simple line buffer example (Example 7-16) is presented below to better understand why this can be a problem.

Example 7-16. The Problem with Using Singleport RAM

```
#include <ac_int.h>
void  test_sp_orig(ac_int<10,false>  din[720],  ac_int
<10,false> dout[720], bool write){
  static ac_int<10,false> ram[720];
  static bool dummy = ac::init_array<AC_VAL_DC>(ram,720);
  for(int i=0;i<720;i++){
    dout[i] = ram[i];
    if(write)
      ram[i] = din[i];
  }
}

Design Constraints
"din", dout, and ram mapped to singleport memory
All loops left rolled
Main loop pipelined with II=1
```

Example 7-16 fails to schedule due to a pipelining failure. In the design the "ram" array mapped to a singleport RAM is always read inside of the loop, and can also be written in the same loop iteration. Pipelining with II=1 would mean that both the read and the write would have to be scheduled in the same clock cycle. Figure 7-5 shows the "failed" schedule for this design. A singleport RAM usually has a read data port, a write data port, and a single address port. Each memory operation in the schedule requires its own address. Thus it is not possible to simultaneously address a singleport RAM for reading and writing.

Figure 7-5. Failed Schedule for Reading and Writing a Singleport RAM with II=1

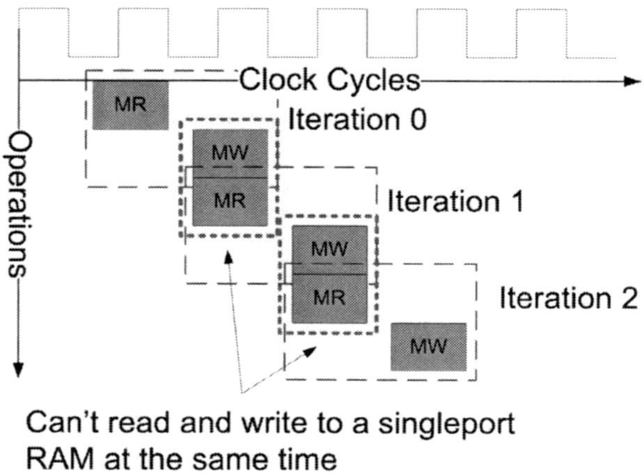

The solution to being able to simultaneously read and write to a singleport RAM and pipeline with II=1 is to write the C++ code so that reads and writes are cached, and forced into different loop iterations or clock cycles. This also requires changing the word width of the array mapped to memory. Using a C++ class allows most of these changes to be encapsulated so that the original code looks mostly the same. Example 7-17 shows an implementation of a class that accomplishes this. Like many of the other examples covered previously, this class assumes that the array accesses consistently start on an even word boundary and then read the next sequential location which would be on an odd word boundary. This class is restricted to using Algorithmic C++ integer types to easily allow manipulation of the word width.

Example 7-17. Singleport RAM Class that Supports II=1

```
#ifndef __SINGLEPORT__
#define __SINGLEPORT__
#include <ac_int.h>
template<int N, int W, bool S>
class singleport_ram{
  ac_int<ac::log2_ceil<N>::val ,false> addr_int;
```

```
7     ac_int<W*2,false> ram[N/2];
8     ac_int<1,false> cnt;
9     ac_int<W*2,false> read_data;
10    ac_int<W*2,false> write_data;
11  public:
12    singleport_ram():cnt(0),read_data(0),write_data(0){
13      bool dummy = ac::init_array<AC_VAL_DC>(ram,N/2);
14    }
15  #include "exec.hpp"
16  };
17  #endif
```

The details of Example 7-17 are:

- Line 4 defines the template parameters for number of array elements "N", word width "W", and signedness of the base type.
- Line 6 uses log2_ceil to ensure that the internal address width is reduced to the minimum number of bits.
- Line 7 defines an internal array "ram" that is "W*2" wide and "N/2" elements. In other words the array is twice as wide and half as many elements. Note that this implementation expects "N" to be evenly divisible by two. Support for an odd number of array elements would require more complex control in the C++.
- Line 8 defines a single bit counter that is used to control reading and writing of data.
- Lines 9 and 10 define two variables that are width "W*2" and are used to perform the internal caching.
- Lines 12 and 13 initialize the counters and use the ac::init_array function to remove the initialization of the "ram" array that 's mapped to memory.
- Line 15 includes "exec.hpp" which implements the read/write method for the class. Although this is not considered the best C++ style it was done simply to allow the code to be discussed in smaller fragments. The code defined in "exec.hpp" is simply inlined where it is included in the class definition.

Example 7-18. Read/Write Method for Singleport RAM Class

```
ac_int<W,S> exec(ac_int<W,S> data_in,int addr, bool write){
    ac_int<W,S> tmp;
    addr_int = addr;
    if(write){
      if(cnt==0)
        write_data.set_slc(0,data_in);
```

```
7         else
8           write_data.set_slc(W,data_in);
9       }
10      if(cnt==0){//read on even
11        read_data = ram[addr_int>>1];
12      }
13      else{//write on odd
14        if(write){
15          ram[addr_int>>1] = write_data;
16        }
17      }
18      if(cnt==0)
19        tmp = read_data.template slc<W>(0);
20      else
21        tmp = read_data.template slc<W>(W);
22      ++cnt;
23      return tmp;
24    }
```

Example 7-18 shows the implementation of the singleport ram class method that allows reading and writing to the singleport emory. The details are:

- Line 1 returns the data read from the array as an ac_int<W,S>, which is the same data type as the original data type. The "exec" function takes as its arguments the data to be written "data_in", the read/write address "addr", and a flag "write" to indicate if the write should occur. Thus the array can be read-only if the "write" flag is false.
- Line 3 assigns the address supplied on the interface of "exec" to the data member "addr_int". "addr_int" has been constrained to the minimum number of bits needed to index the memory when the class was defined.
- Lines 4 through 9 are executed when data is written, "write==true". The ac_int write slice method is used to alternate between writing the data to the lower and upper halves of the "write_data" data member. The single bit counter "cnt" is used to decide which half should be written.
- Lines 10 through 17 control whether the internal array "ram" is read or written. Reads occur only on even addresses (cnt ==0) and writes, when "write==true", on odd addressed (cnt==1). The address is adjusted (addr_int>>1) to account for the memory being "N/2" locations and "W*2" wide.

Example 7-19. Using the Singleport RAM Class

```
#include "singleport_ram.hpp"
void test_sp(ac_int<10,false> din[720],
ac_int<10,false> dout[720], bool write){
  static singleport_ram<720,10,false> ram;
  for(int i=0;i<720;i++){
    dout[i] = ram.exec(din[i],i,write);
  }
}
```

"Windowing" of 1-D Data Streams

The previous section on memory-based shift registers illustrated how array/memory access patterns can restrict the performance of a design. The solution was to limit the number of memory reads and writes to match the memory bandwidth. There are many classes of algorithms that are often expressed in a style, that while natural for algorithm development, are not well written from a high-level synthesis quality of results point of view. The key to achieving high quality hardware is to analyze the way data moves through the algorithm, and to express that movement efficiently in the underlying memory architecture. A simple moving average filter is a good example that can illustrate the limitations of a poorly coded memory architecture, as well as show how restructuring the C++ code can lead to optimal hardware.

Pure Algorithmic Description with Poor Memory Architecture

Example 7-20 shows a simple moving average filter that sums three weighted samples from an array on the design interface that is mapped to memory. It's not unreasonable to expect that this type of algorithm, when implemented in hardware, can compute a new value of "dout[i]" every clock cycle. However, as the design schedule shows, the coding style of Example 7-20 limits the design performance due to a memory bottleneck.

Example 7-20. Simple Moving Average with Poor Memory Architecture

```
#include "window_1d.h"

int clip(int i){
  int tmp = i;
  if(tmp < 0)
```

```
      tmp = 0;
    else if(tmp > NUM_WORDS-1)
      tmp = NUM_WORDS-1;
    return tmp;
}

void avg(ac_int<8,false> din[NUM_WORDS],
         ac_int<8,false> dout[NUM_WORDS]){
    const ac_fixed<3,1,false> coeffs[3] = {0.25, 0.5, 0.25};
    ac_fixed<13,11,false> tmp;
  COMP:for(int i=0;i!=NUM_WORDS;i++){
      tmp = din[clip(i-1)]*coeffs[0] + din[i]*coeffs[1]
        + din[clip(i+1)]*coeffs[2];
      dout[i] = tmp.to_int();
    }
}

Design Constraints
No pipelining
"din" and dout mapped to singleport memory
All loops left rolled
```

The approximate schedule for Example 7-20 is shown in Figure 7-6. For clarity the schedule only shows the memory accesses.

Figure 7-6. Schedule for Moving Average with Poor Memory Architecture

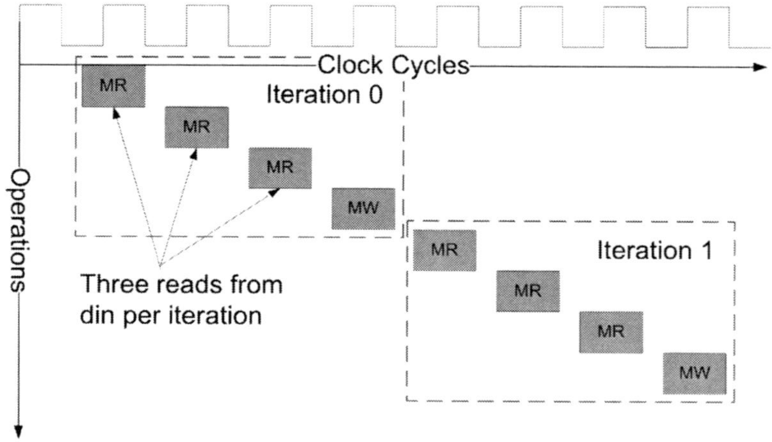

Figure 7-6 shows the un-pipelined schedule for the moving average example. Each loop iteration requires that the "din" memory is read three times. Thus, trying to pipeline this design with II=1 to achieve one value of "dout[i]" per clock cycle is impossible. Figure 7-7 shows the failed schedule when trying to pipeline Example 7-20 with II=1. Since "din" is mapped to a singleport memory only one memory read per clock is possible. However trying to pipeline with II=1 would mean overlapping the loop iterations such that three reads per clock cycle are required, which is impossible. This illustrates how an algorithmic coding style may not have sufficient architectural detail to realize good quality hardware. The next step that should be taken is to analyze the array access patterns of the algorithm to see if the C++ code can be restructured to achieve the desired performance.

Figure 7-7. Failed Schedule for Moving Average with II=1

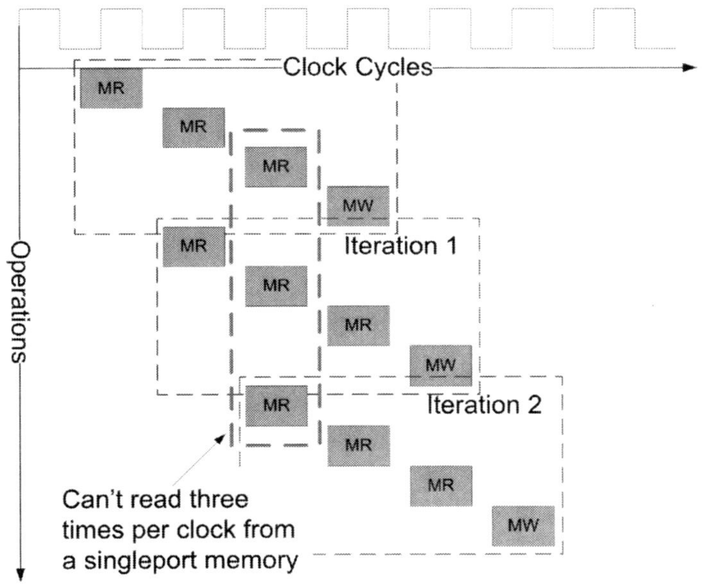

Analyzing Array Access Patterns

Figure 7-8 shows the general access pattern for the moving average algorithm. By writing out a few of the loop iterations a general pattern appears. Each loop iteration reads three values from "din", but the computation of "dout[i]" always uses two values of "din[i]" that were read in the previous iteration. In other words, only one new value of "din[i]" needs to be read for each iteration, and the other values that were previously read can be reused. This implies that storage is required. The reason why this is referred to as windowing is because only a small portion of "din" is required for processing each loop iteration. That small portion of "din" can be thought of as a window that slides over the entire array. This sliding window behavior is easily mapped to a shift register. Since the first loop iteration cannot reuse old values of "din" it needs special handling to account for the "startup" of the hardware as well as the boundary condition defined by the "clip" function".

Figure 7-8. Moving Average Access Pattern

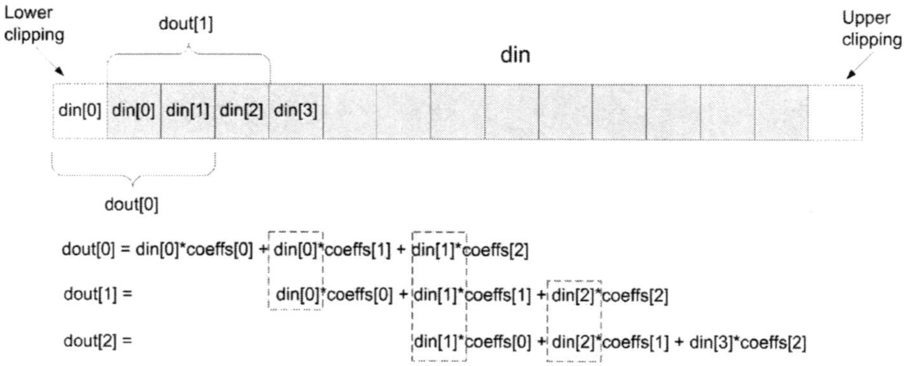

Shift Register Sliding Window Implementation

The simple moving average example discussed above can be re-written to take advantage of the array access patterns in order to reduce the memory access of "din" to once per clock cycle. Example 7-21 shows the moving average filter that is written using a sliding window. There are two principal components, the shift register to store previous values of "din", and the clipping function to handle the boundary conditions.

Example 7-21. Sliding Window Moving Average Filter

```
#include "window_1d.h"
#include "shift_class.h"
void window_avg(ac_int<8,false> din[NUM_WORDS],
          ac_int<8,false> dout[NUM_WORDS]){
  const ac_fixed<3,1,false> coeffs[3] = {0.25, 0.5, 0.25};
  shift_class<ac_int<8,false>, 3> shift_reg;
  ac_int<8,false> window[3];
  ac_fixed<13,11,false> mac;
  ac_int<8,false> din_tmp;

 COMP:for(int i=0;i!=NUM_WORDS+1;i++){
    if(i<NUM_WORDS)//prevent overread of din
      din_tmp = din[i];
    shift_reg << din_tmp;
    clip_window(shift_reg,i,window);
    mac = window[0]*coeffs[0] + window[1]*coeffs[1]
```

```
17        + window[2]*coeffs[2];
18      if(i>=1)//startup
19        dout[i-1] = mac.to_int();
20    }
21  }
```

```
Design Constraints
Main loop pipelined with II=1
"din" and dout mapped to singleport memory
Shift register loops fully unrolled, all other loops left
rolled
```

The details of Example 7-21 are:

- Line 6 declares a shift register class variable with three taps. This is used to store the previous values of "din".
- Line 7 declares a three element array "window" that is used to apply the boundary conditions to the sliding shift register.
- Line 11 has added an additional iteration to the "COMP" loop. This is done to handle the requirement that "dout[0]" requires two reads from "din".
- Lines 12 and 13 reads "din[i]" into "din_tmp" and prevents "din" from being over-read since the "COMP" loop runs for NUM_WORDS+1 iterations. Since "din" is only read once per loop iteration it is possible to pipeline the main loop with II=1.
- Line 14 shifts each new value of "din[i]" into the sliding shift register.
- Line 15 passes the shift register to the clipping function which handles the boundary conditions. It returns the windowed and clipped data in "window".This is discussed next.
- Lines 16 and 17 perform the multiple and accumulate, using the "window" array, which has stored current and past values of "din".
- Lines 18 and 19 begin conditionally writing the output "dout" once enough data has been read to compute "dout[0]". The index of "dout" is adjusted to account for the startup iteration so that the first write begins from location zero.

Boundary Conditions

The "clip" function in Example 7-21 processes the sliding window "shift_reg" so that the returned array "window" has the same boundary behavior based on the index "i". This is shown in Example 7-22.

Example 7-22. Clipping Function for Sliding 1-D Window

```
1   #include "window_1d.h"
2   #include "shift_class.h"
3   void clip_window( shift_class<ac_int<8,false>, 3> shift_reg,
4                int i, ac_int<8,false> window[3]){
5
6     window[0] = (i==1) ? shift_reg[1]:shift_reg[2];
7     window[1] = shift_reg[1];
8     window[2] = (i==NUM_WORDS) ? shift_reg[1]:shift_reg[0];
9   }
10
```

The details of Example 7-22 are:

- Line 6 implements the equivalent of "din[clip(i-1)]" from the original design. The shift register stores the newest data in "shift_reg[0]" so when "i" equals one, the shift register has "din[0]" stored in "shift_reg[1]" and "din[1]" stored in "shift_reg[0]". Figure 7-8 showed that clipping the index from going negative has the effect of duplicating the value of "din[0]". When the index "i" equals one the value of "din[0]" stored in "shift_reg[1]" is copied into "window[0]", implementing the lower clipping.
- Line 8 implements upper clipping when the index "i" equals NUM_WORDS. This comparison is adjust to account for the startup of the window.

Instead of clipping the array index like Example 7-20, the clip function for the sliding window implementation uses the index to control a selection MUX which implements the boundary condition. The hardware synthesized form Example 7-22 is shown in Figure 7-9.

Figure 7-9. Clipping Function for 1-D Sliding Window

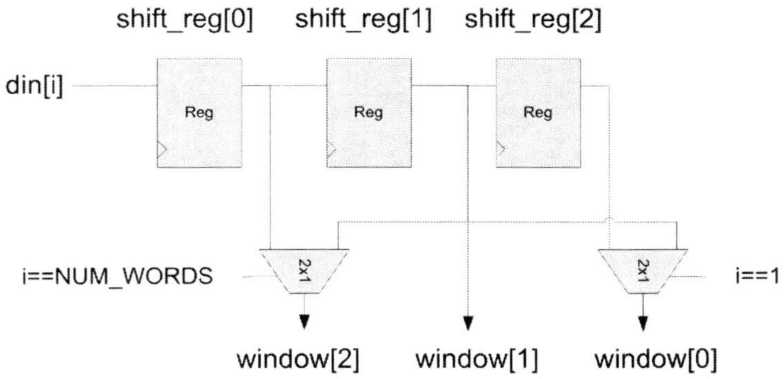

2-D Windowing

1-D windowing showed that a register based memory architecture could be used to cache previous sequential reads from a one dimensional array mapped to memory, allowing designs to run at the maximum output data rate. These same design principles can be applied to two dimensional algorithms, where the implementation of the caching requires a slightly more complicated memory architecture.

Pure Algorithmic Description with Poor Memory Architecture

Example 7-23 shows another moving average filter, but in this case the input is a two dimensional array. The averaging is performed across rows of the array instead of sequential locations like the 1-d moving average example. This example has the same type of bandwidth limitations that were shown in the 1-d example when "din" is mapped to a memory (See ""Windowing" of 1-D Data Streams" on page 179).

Example 7-23. 2-d Moving Average

```
#include "window_2d.h"
int clip(int i){
  int tmp = i;
  if(tmp < 0)
    tmp = 0;
  else if(tmp > 480-1)
    tmp = 480-1;
  return tmp;
}
void avg(ac_int<8,false> din[480][720],
```

```
        ac_int<8,false> dout[480][720]){
    const ac_fixed<3,1,false> coeffs[3] = {0.25, 0.5, 0.25};
    ac_fixed<13,11,false> tmp;

  ROW:for(int r=0;r!=480;r++){
    COL:for(int c=0;c!=720;c++){
            tmp  =  din[clip(r-1)][c]*coeffs[0]  +  din[r][c]*coeffs[1]
      + din[clip(r+1)][c]*coeffs[2];
        dout[r][c] = tmp.to_int();
      }
    }
}
```

Analyzing Array Access Patterns

Figure 7-10 shows how the array accesses occur for Example 7-23.

Figure 7-10. 2-d Moving Average Access Arrays Patterns

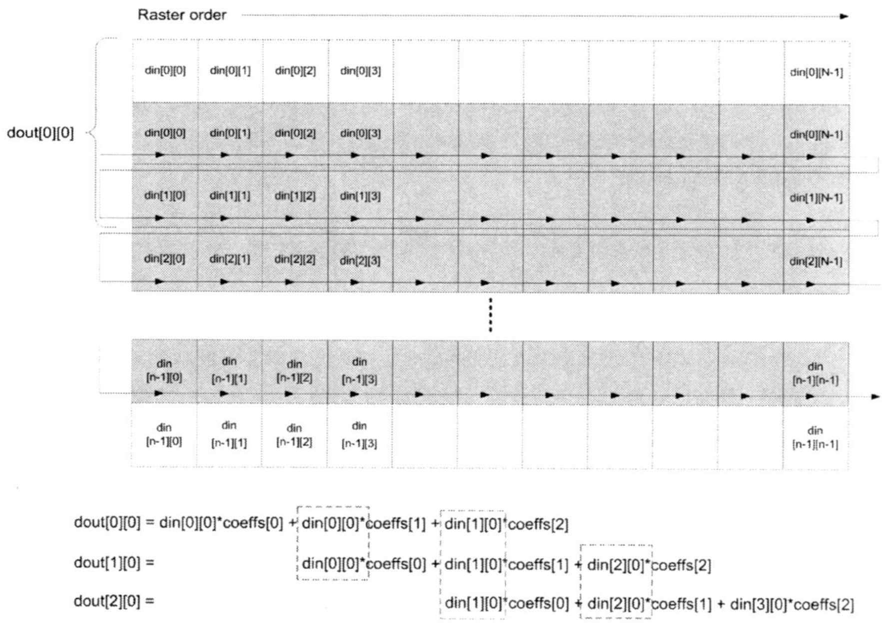

The input array "din" is read in what is known as raster order, which means that the elements of "din" are read sequentially. Each row is read column by column before proceeding to the next row. Although there is no commonality between adjacent columns, writing out a few of the adjacent row computations shows a trend. Excluding the array boundary, the current computation of a row output "dout[r][c]" depends on one new row value and two row values used in the previous row computation. Thus to compute "dout[2][0]" requires reading in one new value "dout[3][0]". The values "dout[1][0]" and "dout[2][0]" were used to compute "dout[1][0] in the previous row computation. This array access pattern shows that only one new value must be read to compute an output if the two previous rows of data are stored locally.

Circular Line Buffer Sliding Window Implementation

The analysis of the array access patterns of the 2-d moving average filter showed that each new row computation uses data read from the previous two rows. If the previous row data is stored internally, it can be reused rather than having to re-fetch it from external memory. Unfortunately using registers for the internal storage, like what was done for the 1-d sliding window, is not practical since each row contains 720 elements. This means that memories are required to store the two rows of data. The most efficient implementation from a power and area standpoint is to use two singleport RAMS to store the previous two rows of data, while using the current row data directly. The RAM buffers are written and read in a circulating fashion to minimize power consumption from switching. The output of the RAMs, along with the current data are then "clipped" in the same fashion as the 1-d window example, where the row index "r" is used to determine when the boundary condition is applied. Figure 7-11 shows the general hardware structure of the implementation.

Figure 7-11. Circular Buffer Window Implementation

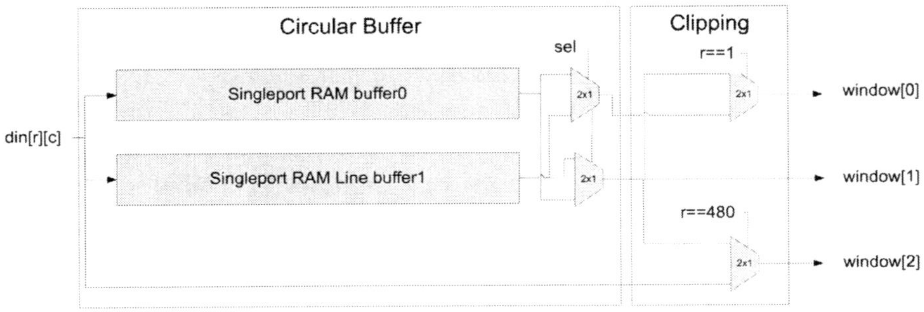

Example 7-24 shows the implementation of the circular buffer portion of the design.

Example 7-24. Circular Buffer for Storing Multiple Rows of Data

```
1   #include "singleport_ram.hpp"
2   #include "window_2d.h"
3   void buffer(ac_int<8,false> din, int c,
4           ac_int<8,false> window[3]){
5     static singleport_ram<720,8,false> buffer0;
6     static singleport_ram<720,8,false> buffer1;
7     ac_int<8,false> b0,b1,b2;
8     ac_int<8,false> t0,t1;
9     static ac_int<1,false> sel=1;
10
11    if(c==0)//switch buffer write at start of line
12      sel += 1;;
13    t1 = buffer1.exec(din,c,sel);
14    t0 = buffer0.exec(din,c,!sel);
15
16    window[0] = (sel==1) ? t1:t0;
17    window[1] = (sel==1) ? t0:t1;
18    window[2] = din;
19  }
```

The details of Example 7-24 are:

- Line 1 includes the singleport RAM class that was described in "Using True Single Port RAM as a Dualport RAM" on page 176.
- Lines 3 and 4 takes the inputs "din", the current address or column position "c", and returns column aligned data from three different rows.
- Lines 5 and 6 instantiate the two line buffers using the singleport RAM class.
- Line 9 defines a one bit counter that is used to select between the two line buffers.
- Line 11 and 12 check the current address or column position "c" and increments "sel" at the start of a new row.
- Lines 13 and 14 call the singleport RAM "exec" function. The input data "din" is passed to both memories along with the address "c". "sel" is only active for one memory at a time so only one of the memories is written. Thus the memories are written in circulating, or alternating, fashion. Both memories are read and return the data in "t1" and "t0"

- Lines 16 through 18 return data from three different rows. The newest data is “din” which is returned directly. The older data stored in “buffer0” and “buffer1” are returned in circulating fashion based on the value of “sel”.

The clipping of the boundary conditions is implemented slightly differently from the 1-d case since there is no shift register class used in this design. However the design principles are the same (Example 7-25).

Example 7-25. Clipping Function for 2-d Sliding Window

```
1   #include <ac_window.h>
2   #include "window_2d.h"
3   void clip_window(int r, ac_int<8,false> window[3]){
4     ac_int<8,false> w[3];
5
6     for(int i = 0;i<3;i++)
7       w[i] = window[i];
8     window[0] = (r==1) ? w[1]:w[0];
9     window[1] = w[1];
10    window[2] = (r==480) ? w[1]:w[2];
11  }
```

The details of Example 7-25 are:

- Lines 2 and 3 - the function takes the input argument “r” the current row index, and “window[3]” the data from the circular buffer. “window[3]” also returns the buffer data with the boundary conditions applied.
- Lines 6 and 7 makes a copy of the original buffer data and stores it in “w”
- Line 8 applies the lower boundary condition when “r==1”. The reason why it is “r==1” instead of “r==0” is because the data must be read sequentially. Thus the first two rows must be read “r==0” and “r==1” before there is enough data to start computing the output.
- Line 10 applies the upper boundary condition when “r==480”.

Example 7-26 shows the rewritten 2-d averaging filter using the circular buffer and clipping function.

Example 7-26. 2-d Moving Average Using 2-d Windowing

```
1   #include "window_2d.h"
2   void window_avg(ac_int<8,false> din[480][720],
3                   ac_int<8,false> dout[480][720]){
4     const ac_fixed<3,1,false> coeffs[3] = {0.25, 0.5, 0.25};
5     ac_fixed<13,11,false> tmp;
6     ac_int<8,false> w[3];
7     ac_int<8,false> din_tmp;
8    ROW:for(int r=0;r!=480+1;r++){
9     COL:for(int c=0;c!=720;c++){
10        if(r != 480)
11          din_tmp = din[r][c];
12        buffer(din_tmp,c,w);
13        clip_window(r,w);
14            tmp = w[0]*coeffs[0] + w[1]*coeffs[1] +
w[2]*coeffs[2];
15        if(r!=0)
16          dout[r-1][c] = tmp.to_int();
17      }
18    }
19  }
```

The details of Example 7-26 are:

- Line 8 extends the ROW loop by one iteration. This is required since the first two rows must be read before the output can start being written.
- Lines 10 and 11 read the input "din" until the last, or extra, iteration of the ROW loop. When on the last ROW iteration "r==480" the reads are disable so that the array is not over-read.
- Line 12 calls the circulating buffer and passes the current data "din_tmp", the column address "c", and the storage for the windowed row data "w".
- Line 13 calls the hardware clipping function. It passes the current row index "r" and the data from the circular buffer "w". "w" is returned with the boundary conditions applied.
- Line 14 uses the windowed data "w" to perform the filter computation.
- Lines 15 and 16 write the output data. The data is not written until the second ROW iteration is in progress "r!=0". The output row index is adjusted to account for this startup delay "dout[r-1][c]".

Chapter 8
Hierarchical Design

Introduction

Up until this point the focus of this book has been on simple block level design. Most of the blocks that were discussed consisted of a single set of loops, or sequential loops, that were completely unrolled. The reality is that many hardware designers are required to build systems, or sub-systems, that contain multiple blocks that run concurrently to one another. Adding this type of concurrency can be done by applying HLS constraints in combination with a recommended coding style, while still leaving the C++ untimed and single threaded. The synchronization of data flow between blocks is added automatically during the synthesis process.

Arrays Shared Between Blocks

One of the most common needs for explicit hierarchy is when two or more functions read and write the same array. These functions typically contain loops that cannot be merged automatically, and manual merging of the loops is time consuming as well as unnatural. Although user-defined hierarchy is most often required as design or control complexity increases, it can easily be illustrated here using very simple, yet somewhat contrived, examples.

Out-of-order Array Access

Consider the following two-block design, shown in Example 8-1, where an array is copied between two blocks. The design is very simple and consists of two blocks “BLOCK0” and “BLOCK1” which copy the top level input array “din” to the output “dout”. The only difference between the two blocks in this design is that the array indexing is done in the opposite order. “BLOCK0” indexes the arrays in ascending order, and “BLOCK1” indexes the arrays in descending order. Although this is a trivial example it illustrates why explicit hierarchy is needed.

Example 8-1. Out of Order Array Access

```
void BLOCK0(int din[3],int dout[3]){
 WRITE:for(int i=0;i<3;i++){
    dout[i] = din[i];
   }
}
void BLOCK1(int din[3],int dout[3]){
 READ:for(int i=2;i>=0;i--){
    dout[i] = din[i];
   }
}
void top(int din[3],int dout[3]){
  int tmp[3];
  BLOCK0(din,tmp);
  BLOCK1(tmp,dout);
}
```

Example 8-1 is synthesized as a flat design with the following constraints:

```
Design Constraints
Main loop pipelined with II=1
All loops left rolled
All arrays mapped to registers
```

The “WRITE” and “READ” loops in functions “BLOCK0” and “BLOCK1” are not automatically merged because the respective array indexing is in the opposite order. Figure 8-1 shows the approximate scheduling of the “BLOCK0” and BLOCK1” functions. Although the last and first loop iterations of the two blocks are overlapped, the loops still execute sequentially, each one having to wait till the other is finished, limiting the overall throughput of the design.

Figure 8-1. Out of Order Array Accesses in a Flat Design

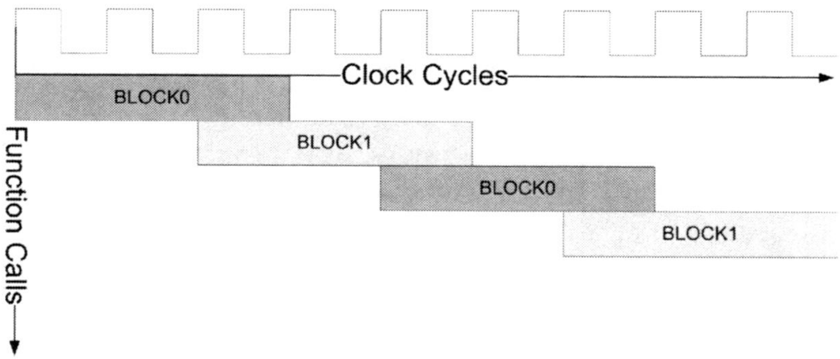

Arrays Mapped to Registers

The throughput limitation shown in Figure 8-1 can be overcome by constraining high-level synthesis to insert user defined hierarchy.

Note

There are a number of ways to define explicit hierarchy, the most common being the application of synthesis constraints at the function level.

When C++ functions are constrained to be hierarchical, they are synthesized as individual blocks, with communications channels between the blocks automatically created based on the arrays and/or variables passed between the functions. The implementation of the communications channels depends on not only whether the design was constrained to use registers or memories, but also on the order of array accesses as well as the read/write control complexity.

Example 8-1 is re-synthesized with the following constraints:

```
Design Constraints
BLOCK0 and BLOCK1 mapped to hierarchy
BLOCK0 and BLOCK1 pipelined with II=1
All loops left rolled
All arrays and channels mapped to registers
```

Synthesizing the design with the constraints given above results in the hardware shown in Figure 8-2. "BLOCK0" makes a local copy of "tmp", the shared array, and writes the entire copy. The array copy is then written from "BLOCK0" to the channel FIFO, which is wide enough to pass the entire array in parallel. "BLOCK1" reads the entire array in parallel from the FIFO and makes a local copy which is then used internally. The FIFO is sized automatically during synthesis but can be overridden with a user defined size. In this example, since there is no handshake on the output (BLOCK1 cannot be stalled) the FIFO size could be set to zero. If the output did have a handshake (BLOCK1 can be stalled) the FIFO can be sized by the user to the appropriate value to prevent loss of data. The FIFO flags are connected to BLOCK0 and BLOCK1 to control both startup as well as stalling behavior. These flags are transparent to the user and are inserted to enforce that the hardware behavior matches the C++ exactly.

Figure 8-2. Out of Order Array Accesses Using Hierarchy and Registers

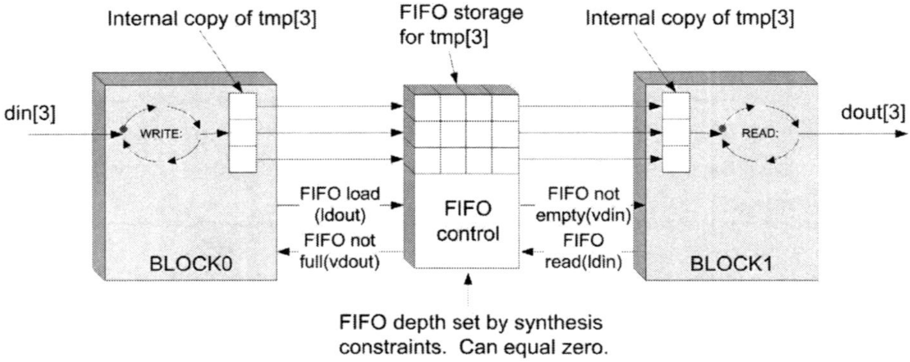

Figure 8-3 shows both the schedule and FIFO flag timing for Example 8-1 when using hierarchy. BLOCK0 runs for three clock cycles after which the FIFO is loaded with the output from BLOCK0. In this example the FIFO and control logic were synthesized with a combinational path through the FIFO, so data is available immediately. This allows BLOCK1 to begin processing immediately. More importantly the schedule shows that the next call of BLOCK0 can begin running while BLOCK1 is processing. Thus the design can run with a continuous throughput after an initial startup latency.

Note

This example uses a FIFO with combinational control for zero latency response to illustrate one aspect of hierarchy. The FIFO control behavior can be made sequential via synthesis constraints.

Figure 8-3. Schedule for Out of Order Array Accesses Using Hierarchy

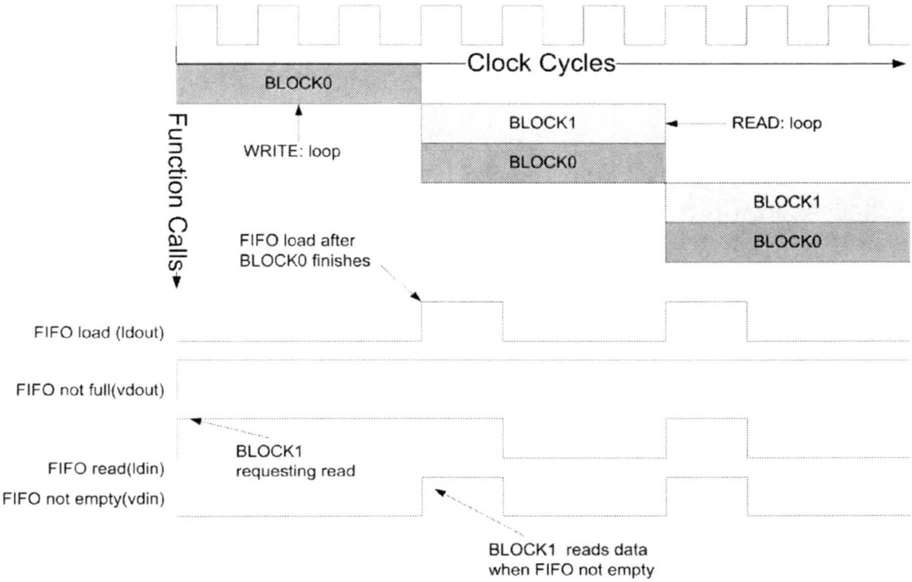

What was shown in this section is that a design with rolled loops that can't be merged can run with maximum throughput by using explicit hierarchy to synthesize functions as separate modules. The control of data flow between the blocks or modules is handled automatically. One possible limitation to this approach would be that as the array sizes of the design become large, the required area for the temporary storage and FIFOs can become excessive. In these cases it is more likely that a memory based hierarchical design may be needed.

Arrays Mapped to Memories

As array sizes become large it is typical to map them to memories during synthesis because the area and power costs of mapping to registers becomes prohibitive. Mapping shared arrays between blocks to memories in a hierarchical design results in a ping-pong memory structure automatically inferred to implement the communications channel. This ping-pong memory structure can consist of two or more memories that are written and read in such an order as to allow the blocks to run concurrently.

Example 8-1 is re-synthesized with the following constraints:

```
Design Constraints
BLOCK0 and BLOCK1 mapped to hierarchy
BLOCK0 and BLOCK1 pipelined with II=1
All loops left rolled
All arrays and channels mapped to singleport RAM
```

Synthesizing the design results in the hardware shown in Figure 8-4. In this case the array "tmp" shared between BLOCK0 and BLOCK1 is stored in a ping-pong memory. The "sync" signal is created to initially stall BLOCK1 until RAM0 is completely written. Once RAM0 is written BLOCK1 can begin reading while BLOCK0 begins writing RAM1. The ping-pong control is generated by BLOCK0 and controls which RAM is written and which is read. The schedule for this ping-pong memory design is identical to the schedule shown in Figure 8-3.

Figure 8-4. Out of Order Array Accesses Using Hierarchy and Memories

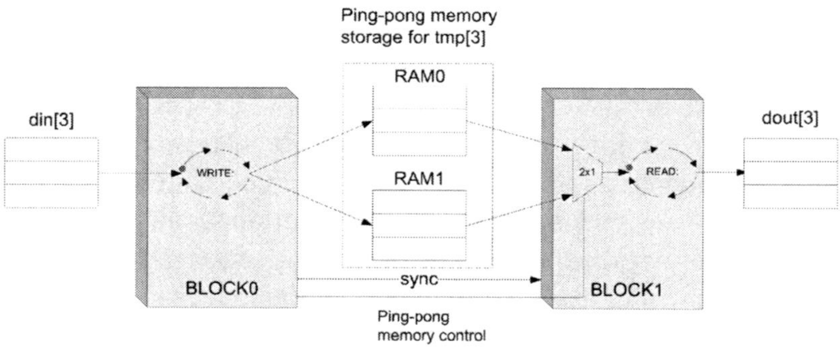

One of the drawbacks to a ping-pong memory hierarchical design is that memories are costly in terms of area. One memory is bad enough, but two is often unacceptable. There are some algorithms, such as the Discrete Cosine Transform (DCT), that often require such an out-of-order memory architecture, but many algorithms can be expressed such that the data transfer between blocks is in the same order.

In-order Array Access

The hardware complexity of the communications channel implementation can be greatly simplified if high-level synthesis can prove that the shared array between blocks is accessed in the same order. The reason why the requirement of proof is highlighted in bold is an important point to understand. In general HLS attempts to build hardware that has a one-to-one IO relationship to the underlying C++ algorithm. This is done so that the RTL generated from synthesis always behaves the same as the C++ algorithm. If this was not enforced people would either think that the RTL was in error, or else they would spend a great deal of time trying to understand why the results were different. Because of this HLS must PROVE that any transformation does not change the functionality. When dealing with shared arrays between blocks this means proving that the indexing of the arrays is the same.

Automatic Streaming

Consider the following simple example that makes a small modification to Example 8-1 on page 192 so that the array indexing is performed in the same order by BLOCK0 and BLOCK1.

Example 8-2. In-Order Array Access

```
void BLOCK0(int din[3],int dout[3]){
 WRITE:for(int i=0;i<3;i++){
    dout[i] = din[i];
  }
}
void BLOCK1(int din[3],int dout[3]){
 READ:for(int i=0;i<3;i++){
    dout[i] = din[i];
  }
}
void top(int din[3],int dout[3]){
  int tmp[3];
  BLOCK0(din,tmp);
  BLOCK1(tmp,dout);
}

Design Constraints
BLOCK0 and BLOCK1 mapped to hierarchy
BLOCK0 and BLOCK1 pipelined with II=1
All loops left rolled
All arrays and channels mapped to registers
```

In Example 8-2 HLS can prove that BLOCK1 reads "tmp" in the same order it's written by BLOCK0. This means that the values written to "tmp" by BLOCK0 do not have to be passed all at once. Rather they can be "streamed" or sent one at a time. Figure 8-5 shows the approximate schedule and FIFO flags for Example 8-2. The values of "tmp" are written one at a time by BLOCK0. BLOCK1 is able to begin reading "tmp" one at a time immediately after the first write by BLOCK0. BLOCK1 stalls on the first clock cycle before the channel FIFO is written, after which it can run every clock cycle.

Figure 8-5. Schedule for In-Order Array Accesses Using Hierarchy

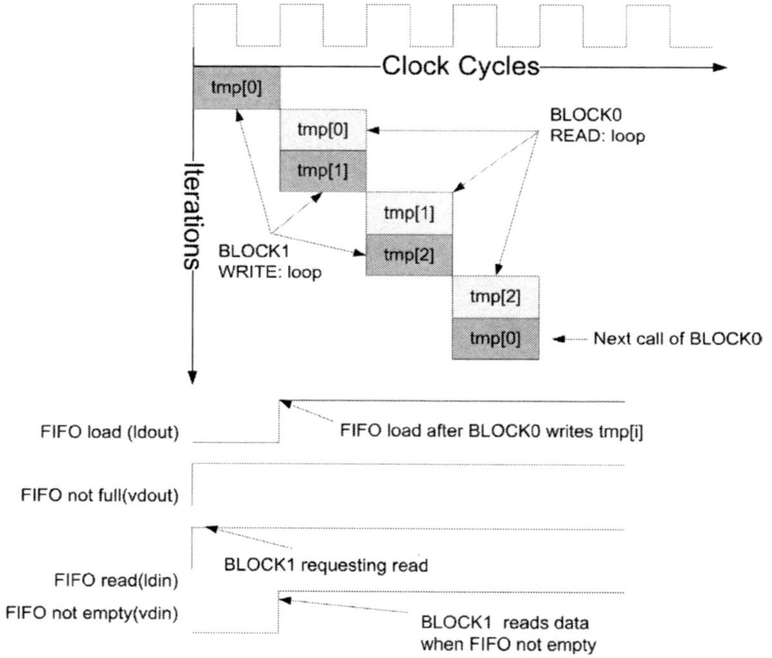

The hardware for Example 8-2 is shown in Figure 8-6. Because the array can be "streamed" the blocks do not have to make internal copies of "tmp", reducing the area. Furthermore the FIFO width needs only to be one element wide, reducing area. The FIFO depth is controllable via synthesis constraints, and can be removed completely if desired.

Figure 8-6. Hardware for In-Order Array Accesses Using Hierarchy

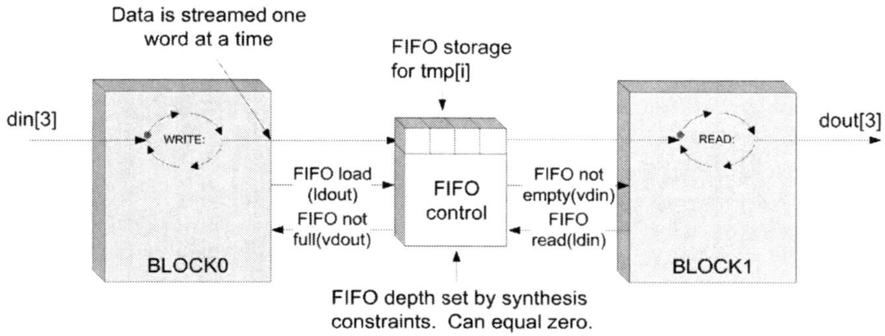

Algorithmic C Channel Class

Caution

Automatic streaming of arrays shared between blocks happens when the array indexing is in the same order and the index computation is unconditional. If these conditions are not met, FIFOs or ping-pong memories are used, leading to larger area.

To put it another way, automatic streaming occurs when the data rates are matched between blocks. Unfortunately there are many classes of algorithms, such as decimation and interpolation, where the array indexing is in the same order, but the index computation has to be conditional. For these class of algorithms HLS can't stream the arrays because it can't prove that doing so would be functionally equivalent. Example 8-3 shows a design that is not streamed. The conditional increment of the index "idx" on line 6 of BLOCK1 prevents streaming. This design would be synthesized with both blocks making copies of the "tmp" array, which would then be passed from BLOCK0 to BLOCK1, resulting in an area-inefficient design.

Example 8-3. Design that Breaks Streaming Between Blocks

```
1   void BLOCK0(int din[4],int dout[2]){
2     static int idx;
3    WRITE:for(int i=0;i<4;i++){
4       if((i&1)==0){
5         dout[idx] = din[i]+din[i+1];
6         idx++;//conditional index modification
7         if(idx==2)
8           idx = 0;
```

```
      }
    }
}
void BLOCK1(int din[2],int dout[2]){
 READ:for(int i=0;i<2;i++){
     dout[i] = din[i];
   }
}
void top(int din[4],int dout[2]){
  int tmp[2];
  BLOCK0(din,tmp);
  BLOCK1(tmp,dout);
}
```

For designs that cannot be automatically streamed, the streaming behavior can be coded directly into the algorithm using the Algorithmic C channel class. The reference manual for the Algorithmic C channel class is available as part of the Catapult Synthesis reference manuals.

Note

The ac_channel class is essentially a C++ FIFO that guarantees that the reading and writing of data between blocks occurs in the same order.

Similar to the coverage of Algorithmic C data types, this book attempts to provide enough of an introduction to AC channels in order to begin writing good quality synthesizable algorithms.

The Algorithmic C Channel class library is included using the following statement:

```
#include <ac_channel.h>
```

Declaration

The ac_channel class is a templatized class which allows specification of the data type passed through the channel. It is declared as:

```
ac_channel<T > my_channel;
```

Where T can be any native, Algorithmic C, SystemC, or user defined data type. For example:

```
ac_channel<ac_fixed<12,6> > my_channel;
```

Note

You must put a space between any two ">" characters or you get a compiler error because the parser treats ">>" as a right shift operator.

Channel Read: T read()

Data is read from the channel using the "read" member function. The member function returns valid data from the channel/FIFO as long as it is not empty. The C++ simulation asserts if the channel/FIFO is read when empty. The synthesized hardware will "block" when attempting to read an empty FIFO. This results in a potential stall of the entire design and is covered in more detail later.

Example:

```
ac_fixed<12,6> tmp = my_channel.read();
```

Note

Channel reads are considered to be a blocking read. What this means is that the reads from a channel must match the writes to a channel or else the system may stall. If a read was allowed on the channel FIFO when it is empty it would underflow, invalidating the data produced by the system. Because of this the read must be "blocked" until data is available in the channel.

Channel Write: write(T)

Data is written to the channel using the "write" member function. The C++ simulation asserts if the FIFO is written when full. The synthesized hardware will "block" when attempting to write a full FIFO. This also results in a potential stall of the entire design and is covered in more detail later.

Example:

```
my_channel.write(tmp);
```

Channel data available: available(int N)

The "available" member function is used to prevent reading an empty channel during C++ simulation, which causes an assertion. It is synthesized as a handshake in hardware so care must be taken when using it since there is the potential for stalling or deadlocking a design. The next chapter covers designs requiring "reactive control", where something else must be done when data is not available.

Example:

```
void test(ac_channel<int> din, ac_channel<int> dout){
if(din.available()){//Hardware stalls until data ready
   int tmp =din.read();
```

Using Explicit Channels

Example 8-3 showed a simple case where the conditional increment of the array index prevents streaming. The example can easily be rewritten using ac_channel to enforce the streaming behavior. This is shown in Example 8-4

Example 8-4. Using ac_channel to Enforce Streaming Between Blocks

```
#include <ac_channel.h>
void BLOCK0(int din[4],ac_channel<int> &dout){
  static int idx;
 WRITE:for(int i=0;i<4;i++){
    if((i&1)==0)
      dout.write(din[i]+din[i+1]);
  }
}
void BLOCK1(ac_channel<int> &din,int dout[2]){
 READ:for(int i=0;i<2;i++){
    dout[i] = din.read();
  }
}
void top_chan(int din[4],int dout[2]){
  static ac_channel<int> tmp;
  BLOCK0(din,tmp);
  BLOCK1(tmp,dout);
}
```

```
Design Constraints
BLOCK0 and BLOCK1 mapped to hierarchy
BLOCK0 and BLOCK1 pipelined with II=1
All loops left rolled
All arrays and channels mapped to registers
```

The details of Example 8-4 are:

- Line 1 includes the ac_channel class library.
- Line 2 has changed the “dout” interface variable of BLOCK0 from an array to a channel. It should be noted that “dout” is declared as a reference because it is an output.
- Line 6 performs a channel write into “dout”.
- Line 9 has changed the BLOCK1 interface variable “din” from an array to a channel. It should be noted that “dout” is declared as a reference. This is required because passing by value would imply making an internal copy, which would lead to larger area.
- Line 11 performs a channel read.
- Line 15 defines a channel at the top level design which is used to connect BLOCK0 and BLOCK1. It must be declared static so that data stored in the channel persists between calls to the top-level design.

Synthesizing Example 8-4 gives a similar schedule and hardware to those shown in Figure 8-5 and Figure 8-6 on page 197.

Note

Channels must always be declared as references when used on a function interface.

Note

Channels must be declared as static when used at the top-level design to connect blocks together.

Using Channels at the Top-level Interface and Testbench

In general transforming a design to use ac_channel is straightforward. Typically ac_channel is used to replace arrays when the array access patterns reflect “streaming” behavior. In other words the array data is written and read in the same order. The general rules for transforming designs and test benches to use ac_channel are:

- Arrays must be read and written in the same order.
- The data rates of reading and writing the array must match. Otherwise the system stalls.
- Array accesses should be replaced with channel reads and writes.
- ac_channel used as formal arguments on functions must be declared using references.
- ac_channel used as interconnections between hierarchical blocks must be declared as static variables.
- ac_channel read by the top-level design must be pre-loaded with data in the C++ testbench before calling the function. Otherwise the C++ simulation asserts.

The following simple design and testbench show an array based design that reads and multiplies two arrays in order and writes the result to an output array.

Example 8-5. Simple Array Based Design

```
void top(int din0[32], int din1[32], int dout[32]){
  for(int i=0;i<32;i++){
    dout[i] = din0[i]*din1[i];
  }
}

```

Example 8-6. Testbench for Simple Array Based Design

```
#include <stdio.h>
#include <stdlib.h>
#include “top_array.h”
int main(){
  int din0[32];
  int din1[32];
  int dout[32];

  for(int i=0;i<32;i++){
    din0[i] = rand();
    din1[i] = rand();
  }
  top(din0,din1,dout);
  for(int i=0;i<32;i++)
    printf(“dout[%d] = %d\n”,i,dout[i]);
  return 0;
}
```

The testbench in Example 8-6 initializes the 32 element arrays “din0” and “din1” with data which are then passed to the function “top”. The function, shown in Example 8-5, reads all 32 elements of “din0” and “din1” in order, multiplies them and writes all 32 elements of “dout” in order. This behavior must be replicated when converting to channels.

The examples shown below show the design and testbench of Examples 8-5 and 8-6 transformed to use ac_channel.

Example 8-7. Simple Array based Design Transformed to use Channels

```
1   #include <ac_channel.h>
2   void top(ac_channel<int > &din0,
3            ac_channel<int > &din1,
4            ac_channel<int > &dout){
5     for(int i=0;i<32;i++){
6       int tmp = din0.read()*din1.read();
7       dout.write(tmp);
8     }
9   }
```

The details of Example 8-7 are:

- Lines 2 through 4 have converted the interface arrays from Example 8-5 to ac_channel. References are use for the declarations.
- Line 6 has replaced the array read accesses with channel reads.
- Line 7 has replaced the array write accesses with channel writes.

Example 8-8. Testbench Using Channels

```
#include <stdio.h>
#include <stdlib.h>
#include “top_channel.h”
#include <ac_channel.h>
int main(){
  ac_channel<int > din0;
  ac_channel<int > din1;
  ac_channel<int > dout;
  for(int i=0;i<32;i++){
    din0.write(rand());
    din1.write(rand());
  }
  top(din0,din1,dout);
  for(int i=0;i<32;i++)
    printf(“dout:%d = %d\n”,i,dout.read());
  return 0;
}
```

The details of Example 8-8 are:

- Lines 6 through 8 - The testbench arrays have been converted to ac_channel.
- Lines 9 through 12 - instead of initializing the elements of an array, the channels read by the top-level function are written with the initialization data.
- Line 15 - the output from the top level design must be read out of the channel one element at a time.

Blocks with Common Interface Control Variables

Multi-block designs often use top-level interface variables to configure the different blocks in the system. It is not uncommon to have a single variable control the behavior of multiple blocks. A good example of this would be a control variable that sets the maximum number of loop iterations for each block. The style in which the control should be written into the C++ depends on the application.

Passing Control Variables Between Blocks

Communications algorithms tend to have the requirement that the system runs without interruption. For these types of designs the control variables should be copied between blocks along with the data to ensure a one to one match between the RTL and the C++. By keeping the control synchronized with the data, upstream changes can be made without having to wait for the system to flush. Example 8-9 illustrates how to pass the control through a channel.

Example 8-9. Passing Control Variables Between Blocks

```
#include "passing_control.h"
void BLOCK0(ac_channel<int> &din,ac_channel<int> &dout,
            ac_channel<ac_int<WIDTH,false> > &ctrl,
            ac_channel<ac_int<WIDTH,false> >&ctrl_out){
  ac_int<WIDTH,false> ctrl_int = ctrl.read();
  ctrl_out.write(ctrl_int);//one write
 WRITE:for(int i=0;i!=ctrl_int;i++){
    dout.write(din.read()*13);
  }
}
void BLOCK1(ac_channel<int> &din,ac_channel<int> &dout,
            ac_channel<ac_int<WIDTH,false> > &ctrl){
  ac_int<WIDTH,false> ctrl_int = ctrl.read();//one read
```

```
14    READ:for(int i=0;i!=ctrl_int;i++){
15       dout.write( din.read());
16     }
17  }
18  void top(ac_channel<int> &din,ac_channel<int> &dout,
19           ac_channel<ac_int<WIDTH,false> > &ctrl){
20    static ac_channel<int> data_int;
21    static ac_channel<ac_int<WIDTH,false> > ctrl_int;
22    BLOCK0(din,data_int,ctrl,ctrl_int);
23    BLOCK1(data_int,dout,ctrl_int);
24  }

Design Constraints
BLOCK0 and BLOCK1 mapped to hierarchy
BLOCK0 and BLOCK1 pipelined with II=1
All loops left rolled
All channels mapped to registers
```

The details of Example 8-9 are:

- Lines 18 and 19 define the top level interface using channels for all interface variables. It is generally a good rule of thumb to not mix channel and non-channel variables in order to avoid deadlocking the system.
- Lines 20 and 21 define two static channels that are used to pass the data and the control between blocks.
- Line 5 reads the BLOCK0 control input "ctrl" and stores it in a temporary variable. This is done so that the control interface is only read once. The temporary variable "ctrl_int" is then used internally to avoid generating multiple read strobes on the top-level control interface. Reading the BLOCK0 control multiple times has the potential to deadlock the system.
- Line 6 writes the control to the BLOCK0 output "ctrl_out" using the internal copy "ctrl_int". It is important to note that this is done outside of the loop to avoid writing the control multiple times. Writing the control to the output of BLOCK0 multiple times also has the potential to deadlock the system if it's not read the same number of times by BLOCK1.
- Lines 7 through 9 copy the BLOCK0 data input channel "din" to the data output channel "dout". The loop is run for "ctrl_int" iterations. The internal copy of the control "ctrl_int" is used to avoid reading the control channel "ctrl" multiple times.
- Line 13 reads the BLOCK1 control channel and stores the result in an internal variable "ctrl_int". This single read of the control channel matches the single write of the channel in BLOCK0. If

the number reads and the writes did not match the system would deadlock due to FIFO overflow or underflow.
- Lines 14 through 16 copies the BLOCK1 input data channel "din" to the output data channel "dout". The number of loop iterations is controlled by the internal copy of the control channel "ctrl_int".

Figure 8-9 shows the general hardware structure for Example 8-9. The control variable "ctrl" can be changed dynamically without having to flush the pipeline because it is synchronized with the movement of the data through the pipeline. The drawback of this approach is that it requires larger area since the control storage requires as many registers as there are pipeline stages. Thus this technique should only be used when needed.

Figure 8-7. Passing Control Variables Between Blocks

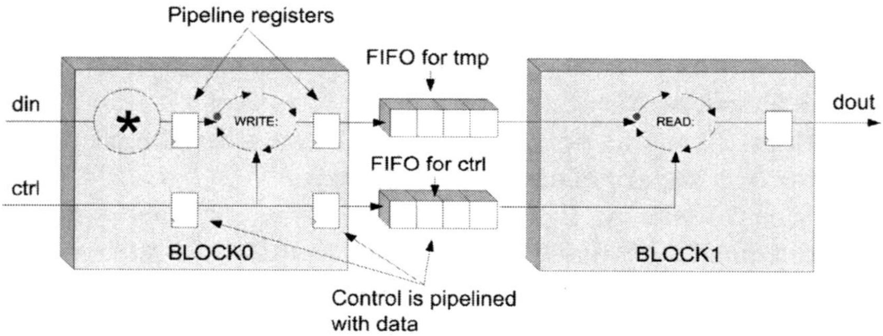

Connecting Interface Control Variables to Multiple Blocks

For video or image processing algorithms data tends to flush completely from the system before the next transaction is processed. This allows plenty of time for changing the control while the system is idle, avoiding the problem of mismatching between the RTL and C++. Driving the control directly from the top-level interface to multiple blocks is typically done for these designs. The control pipeline registers can be bypassed because it does not have to be synchronized to the data. This leads to smaller designs in general. However, the RTL mismatches with the C++ if the control input is changed while the system is running. The mismatches last for the amount of time it takes to flush the old data from the pipeline. If this type of "direct" input is desired, it requires that the control input is defined as a reference at the design interface. Use of an ac_channel does not work since it requires synchronization. Example 8-10 shows how control inputs can simultaneously drive multiple blocks.

Example 8-10. Connecting Interface Control Variables to Multiple Blocks

```
1  #include “direct_input.h”
2  void BLOCK0(ac_channel<int> &din,ac_channel<int> &dout,
3              ac_int<WIDTH,false> &ctrl){
4   WRITE:for(int i=0;i<1024;i++){
5      if(i==ctrl)
6        break;
7      dout.write(din.read()*13);
8    }
9  }
10 void BLOCK1(ac_channel<int> &din,ac_channel<int> &dout,
11             ac_int<WIDTH,false> &ctrl){
12  READ:for(int i=0;i<1024;i++){
13     if(i==ctrl)
14       break;
15     dout.write( din.read());
16   }
17 }
18 void top(ac_channel<int> &din,ac_channel<int> &dout,
19          ac_int<WIDTH,false> &ctrl){
20   static ac_channel<int> data_int;
21   BLOCK0(din,data_int,ctrl);
22   BLOCK1(data_int,dout,ctrl);
23 }
```

```
Design Constraints
BLOCK0 and BLOCK1 mapped to hierarchy
BLOCK0 and BLOCK1 pipelined with II=1
All loops left rolled
All channels mapped to registers
ctrl mapped to a direct input
```

The details of Example 8-10 are:

- Lines 3, 11 and 19 - the control variables “ctrl” is passed on all interfaces as a reference. This is required in order to map it to a direct input. Passing by value would require making a copy, which in turn would create an additional level of hierarchy at the top-level.
- Lines 5 and 13 - “ctrl” can be used directly if it is mapped to a direct input. Direct inputs are considered wire type interfaces and cannot stall the design.

Figure 8-8 show the approximate hardware structure of Example 8-10. By mapping "ctrl" to a direct input it is connected directly to the loop control, bypassing all pipeline register stages. Thus a change on "ctrl" has an instantaneous effect, making all the current data in the pipeline invalid.

Figure 8-8. Connecting Interface Control Variables to Multiple Blocks

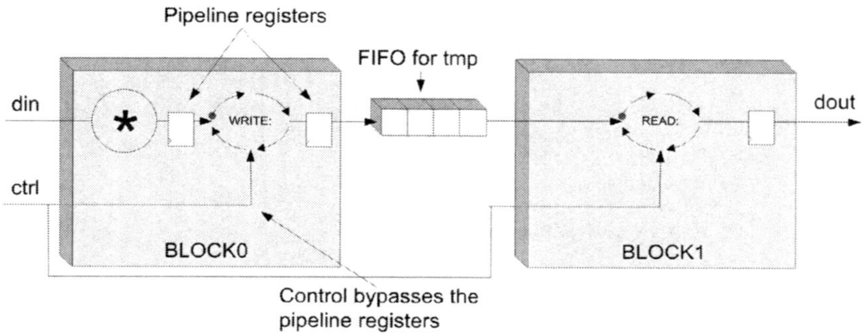

Duplicating Control IO

Some classes of designs require that the control is simultaneously broadcast to multiple blocks, while still remaining synchronized with the data. For these types of designs, it requires making multiple copies of the control signal and then broadcasting the copies. The recommended way of doing this is to create a separate block of hierarchy that makes the control copies and sends them to all blocks. This is shown below in Example 8-11.

```
Design Constraints
CONTROL, BLOCK0 and BLOCK1 mapped to hierarchy
CONTROL, BLOCK0 and BLOCK1 pipelined with II=1
All loops left rolled
All channels mapped to registers
```

The details of Example 8-11 are:

- Lines 2 through 9 implement the control copy block. This function simply reads the control channel "ctrl" once, and then writes it to two output channels "ctrl0" and "ctrl1".

- Lines 12 and 21 - Both BLOCK0 and BLOCK1 read the control channel once, and store it internally in a local variable. The internal copy is then used for the loop control.
- Lines 33 and 34 define two static channels used for copying the control “ctrl”.

Example 8-11. Duplicating Control IO

```
1  #include "duplicate_control.h"
2  void CONTROL(ac_channel<ac_int<WIDTH,false> > &ctrl,
3               ac_channel<ac_int<WIDTH,false> > &ctrl0,
4               ac_channel<ac_int<WIDTH,false> > &ctrl1){
5    ac_int<WIDTH,false> ctrl_int;
6    ctrl_int = ctrl.read();
7    ctrl0.write(ctrl_int);
8    ctrl1.write(ctrl_int);
9  }
10 void BLOCK0(ac_channel<int> &din,ac_channel<int> &dout,
11             ac_channel<ac_int<WIDTH,false> > &ctrl){
12   ac_int<WIDTH,false> ctrl_int = ctrl.read();//one read
13  WRITE:for(int i=0;i<1024;i++){
14     if(i==ctrl_int)
15       break;
16     dout.write(din.read()*13);
17   }
18 }
19 void BLOCK1(ac_channel<int> &din,ac_channel<int> &dout,
20             ac_channel<ac_int<WIDTH,false> > &ctrl){
21   ac_int<WIDTH,false> ctrl_int = ctrl.read();//one read
22  WRITE:for(int i=0;i<1024;i++){
23     if(i==ctrl_int)
24       break;
25     dout.write( din.read()*111);
26   }
27 }
28 void top(ac_channel<int> &din0,
29          ac_channel<int> &din1,
30          ac_channel<int> &dout0,
31          ac_channel<int> &dout1,
32          ac_channel<ac_int<WIDTH,false> > &ctrl){
33   static ac_channel<ac_int<WIDTH,false> > ctrl0;
34   static ac_channel<ac_int<WIDTH,false> > ctrl1;
35   CONTROL(ctrl,ctrl0,ctrl1);
36   BLOCK0(din0,dout0,ctrl0);
37   BLOCK1(din1,dout1,ctrl1);
38 }
```

Figure 8-9 shows the approximate hardware synthesized from Example 8-11. The control is duplicated in the CONTROL block and passed to the rest of the system.

Note

There are no FIFOs shown for the CONTROL block output channels, but these can be added/removed via synthesis constraints.

Figure 8-9. Duplicating Control IO

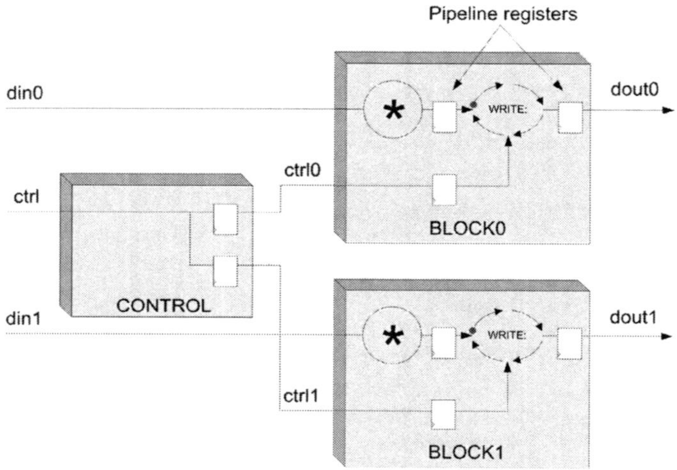

Reconvergence: Balancing the Latency Between Blocks

Many multi-block designs require additional delay registers between blocks to allow the system to run at the maximum throughput. These delay registers, implemented using FIFOs, are used to solve the problem of reconvergence.

Note

Reconvergence occurs when a stream of data travels through blocks with different latencies and is then brought back together.

When the data streams are brought back together they must be aligned in order for the RTL to match the original C++. This alignment is performed by setting the channel FIFO depths large enough so that they can absorb the different latencies between blocks. HLS can automatically size the

FIFOs for simple designs, but designers typically have to choose the appropriate FIFO sizes for complex designs.

Caution

Failing to set the appropriate FIFO size, or setting all FIFOs to zero size, can have the effect of causing the design to either "stutter", where the inputs and outputs are not read/written continuously, or deadlock where the system does not operate.

Deadlock

Consider the following multi-block design shown in Example 8-12. This design is so simple that it really doesn't need user defined hierarchy because there are no loops, but it easily shows the impact of not balancing latency between blocks.

Example 8-12. Multi-Block Design with Reconvergence

```
#include <ac_channel.h>
void BLOCK0(ac_channel<int> &din0, ac_channel<int> &din1,
ac_channel<int> &dout0,ac_channel<int> &dout1){
  int tmp0,tmp1;
    tmp0 = din0.read();
    tmp1 = din1.read();
    dout0.write(tmp0+tmp1);
    dout1.write(tmp0-tmp1);
}
#include <ac_channel.h>
void BLOCK1(ac_channel<int> &din, ac_channel<int> &dout){
  int tmp;
    tmp = din.read()*13;
    dout.write(tmp);
}
void BLOCK2(ac_channel<int> &din1, ac_channel<int> &dout){
  int tmp;
    tmp = din1.read() + 111;
 dout.write(tmp);
}
void BLOCK3(ac_channel<int> &din0, ac_channel<int> &din1,
ac_channel<int> &dout){
  int tmp;
    tmp = din0.read()-din1.read();
    dout.write(tmp);
}
void top_chan(ac_channel<int> &din0, ac_channel<int> &din1,
ac_channel<int> &dout){
  static ac_channel<int> b1_in;
  static ac_channel<int> b2_in;
  static ac_channel<int> b3_in_0;
```

```
    static ac_channel<int> b3_in_1;
    BLOCK0(din0,din1,b1_in,b2_in);
    BLOCK1(b1_in,b3_in_0);
    BLOCK2(b2_in,b3_in_1);
    BLOCK3(b3_in_0,b3_in_1,dout);
}
```

```
Design Constraints
BLOCK0,BLOCK1,BLOCK2, and BLOCK3 mapped to hierarchy
All blocks pipelined with II=1
All channels mapped to registers
All FIFO depts set equal to zero
BLOCK1 uses two-stage pipelined multiplier
```

Figure 8-10 shows the approximate hardware synthesized for Example 8-12. This design deadlocks for the synthesis constraints described above. After the second clock cycle, BLOCK2 is ready to write to BLOCK3. The problem is that the BLOCK2 data must be aligned with the BLOCK1 data. However, BLOCK1 has a longer latency, hence the BLOCK1 data is not available for a few more clock cycles. Since BLOCK2 cannot complete the write it stalls BLOCK0. Since BLOCK0 has to stop writing data to BLOCK2 and BLOCK1, BLOCK1 must also stall, leaving the data stuck in the pipeline. This design never finishes. There are two possible solutions to fix this design, one is two balance the latency between blocks by setting the appropriate FIFO depths, the other is to enable automatic pipeline flushing.

Figure 8-10. Multi-Block Design with Reconvergence

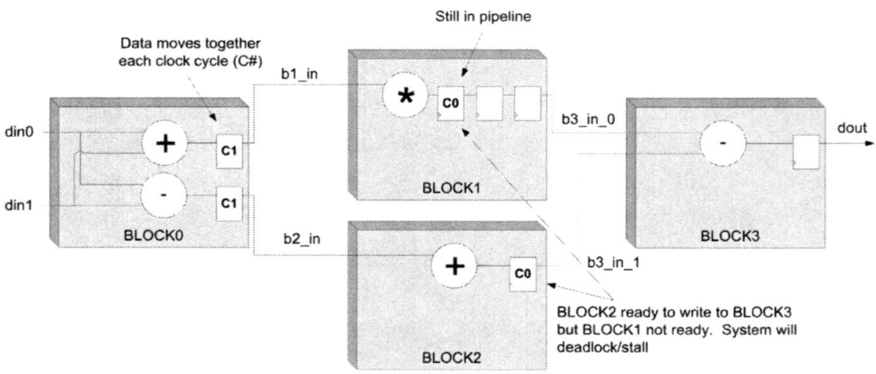

Automatic Pipeline Flushing

The default synthesis flow, because it is the least costly in terms of design area, is to stall the pipeline when data cannot be read or written. This is a result of the ready/acknowledge handshakes that are used for the channel FIFOs. The handshake is connected directly between blocks if the FIFO depth is set equal to zero. Thus if no new data is available at the input, the hardware can stall, leaving data stuck in the pipeline. The pipeline can be made to flush even if there is no new data by enabling automatic pipeline flushing. Figure 8-11 shows the approximate schedule of Example 8-12 when automatic pipeline flushing is enabled.

Figure 8-11. Multi-Block Design with Automatic Pipeline Flushing

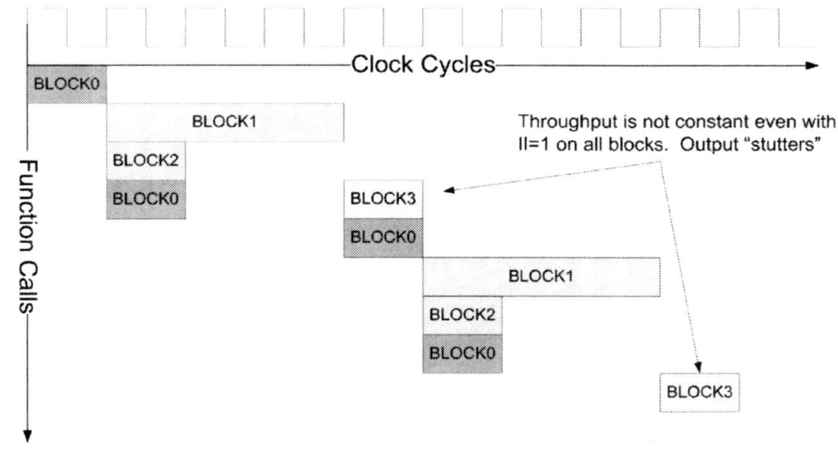

Figure 8-11 shows that even though BLOCK0 and BLOCK2 stall, BLOCK1 is able to complete, allowing BLOCK3 to write the output. The limitation here is that the pipeline is underutilized which can be seen by the blocks starting and stopping continuously.

Manually Setting FIFO Depths

Note

Most reasonably complex design require that the designer manually set the depth of the channel FIFOs to prevent deadlock or stuttering.

Determination of the FIFO depths requires scheduling the design to get the latency information of each block. Once the block latencies are known the designer can assign the appropriate FIFO depths based on how blocks are interconnected. The interconnect information is provided graphically in some HLS tools via block level constraints, or can be determined by inspection. The FIFO depths should be set so that the different latencies in the data stream due to reconvergence are balanced. Figure 8-12 shows Example 8-12 where the FIFO depths have been set to prevent deadlock. Figure 8-10 showed that the reconvergence problem occurred because BLOCK2 was ready to write BLOCK3 two clock cycles before BLOCK1. The deadlock or stuttering can be eliminated by setting the FIFO depth equal to two on the channel between BLOCK2 and BLOCK3.

Figure 8-12. Manually Setting FIFO Depths

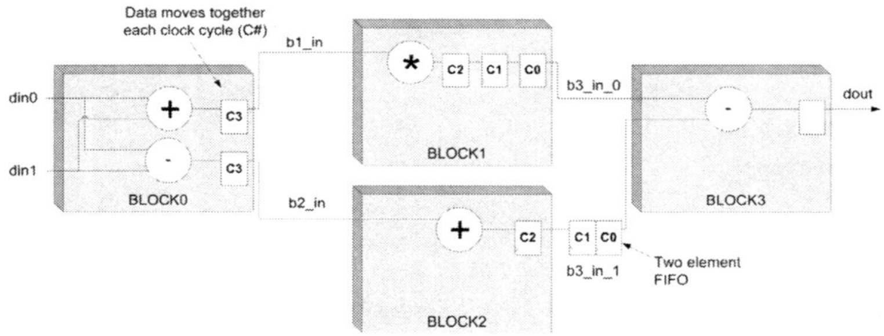

Figure 8-13 shows the effect of setting the FIFO depths to balance latency between blocks. By adding the two element FIFO all of the blocks can start every clock cycle after the initial rampup of the system.

Figure 8-13. Schedule when FIFO Depths Set to Balance Block Latency

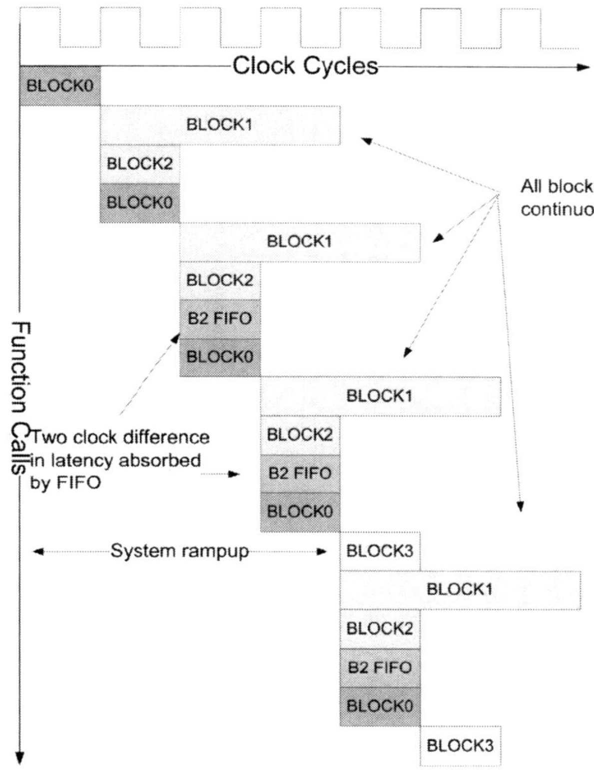

Chapter 9
Advanced Hierarchical Design

Introduction

The previous chapter covered basic hierarchical design and included concepts such as ping-pong memory and communications channel management using the ac_channel class. One of the limitations that was discussed was the potential for stalling the hardware when trying to read an empty channel. There are a number of types of designs where it is essential to be able to "do something else" if data is not available in the channel. Some examples would be memory arbiters, simple bus interfaces, multi-rate designs, and designs with non-deterministic feedback or data flow.

Because of the problems associated with reading empty channels, advanced hierarchical design introduces the concept of a "non-blocking read", which allows transparent accesses to the channel FIFO fullness count. Being able to probe the FIFO size allows designs to be built that never stall the system. Although this is essential when designing the types of designs listed above, it can break the one-to-one relationship between the C++ and generated RTL. Because of this, HLS design environments must support more advanced verification methodologies for proving the correctness of the system, or the resulting hardware must be simulated against the specification instead of the C++ model.

ac_channel Methods

The ac_channel class ("Algorithmic C Channel Class" on page 198) provides two methods to support non-blocking reads.

Channel size: int size()

This member function returns the number of elements in the channel FIFO. A non-blocking read is implemented by using the "size" member function to conditionally read the channel.

Example 9-1. Reading the Channel Size

```
ac_channel<int> input;
int data;
bool flag;
...
flag = input.size()>0;
if(flag)//if data in the FIFO
   data = input.read()
else
   //do something else
```

The read of the channel size is combinational so it can be performed in the same clock cycle as the channel read.

Non-blocking Read: bool nb_read(T &val)

The non-blocking read member function returns true if data is read from the channel. The data is returned in "val". If data is not read from the channel the function returns false.

Example 9-2. Non-blocking Read

```
ac_channel<int> input;
int data;
bool flag;
...
flag = input.nb_read(data);
if(flag)//if data read from FIFO
   //process data
else
   //do something else
```

Recommended Coding Style

Non-blocking reads and the testing of the "size" member function should be used unconditionally when possible. This is because scheduling can't move them out of conditions, which can lead to less than optimal hardware. Consider the following code fragment where the size of different channels is read conditionally

Example 9-3. Bad Coding Style for Reading Channel Size

```
ac_channel<int> input0;
ac_channel<int> input1;
int data;
```

```
...
if(input0.size()>4)
   data = input0.read();
else if(input1.size()>2)
   data = input1.read();
```

Figure 9-1 shows an approximate schedule for Example 9-3. Because the channel size is read conditionally there is a scheduling dependency chain that requires that the channels sizes are read sequentially. This has the potential to increase the latency of the design which can lead to larger area and failure to pipeline.

Figure 9-1. Schedule of Bad Coding Style for Reading Channel Size

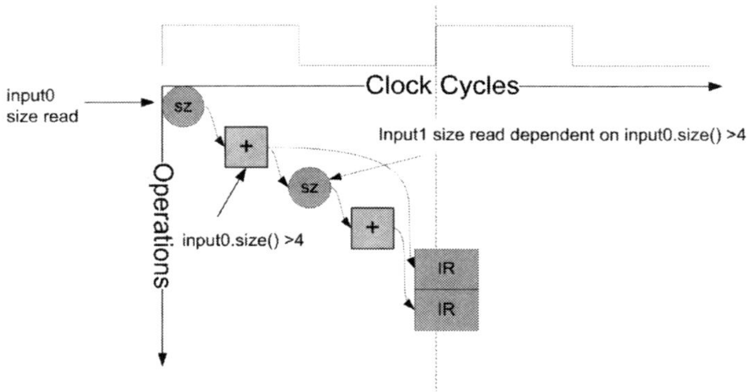

Note

The channel size should always be read unconditionally and stored in a temporary variable.

Example 9-4 shows Example 9-3 rewritten using the recommended coding style. The read of the channel sizes is done unconditionally at the beginning of the design and stored in temporary variables. The temporary variables can then be used directly within the design.

Example 9-4. Recommended Coding Style for Reading Channel Size

```
ac_channel<int> input0;
ac_channel<int> input1;
int data;
bool p[2];
...
p[0] = input0.size()>4;
p[1] = input1.size()>2;
if(p[0])
```

```
      data = input0.read();
   else if(p[1])
      data = input1.read();
```

Figure 9-2 shows the approximate schedule for Example 9-4. By making the reads of the channel sizes unconditional, all of the size comparisons can be done in parallel. The conditional selection logic for the channel reads are then fed with the results of the size comparisons. This logic is minimal and is not shown in the schedule diagram.

Figure 9-2. Recommended Coding Style for Reading Channel Size

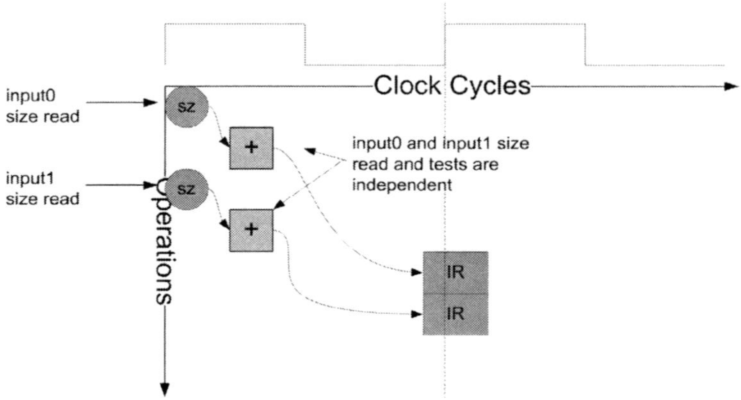

Arbitration

Arbitration is typically required when two or more processes are trying to access the same bus or memory. The bus or memory arbiter processes the requests from the different processes and decides who gets access to the bus/memory. The granting of access can be done based on either priority, or in round-robin fashion where every process is given equal access to the bus/memory. Non-blocking reads are required in order to build things like arbiters. Consider the following simple frame buffer example, shown in Figure 9-3, that consists of two processes accessing the same memory. One process, BLOCK0, writes the memory and the other process, BLOCK1, reads the memory. An arbiter decides who gets access to the memory if both processes are requesting. There is synchronization that controls when, and which half of the memory, the reader can start reading

Figure 9-3. Memory Arbiter Block Diagram

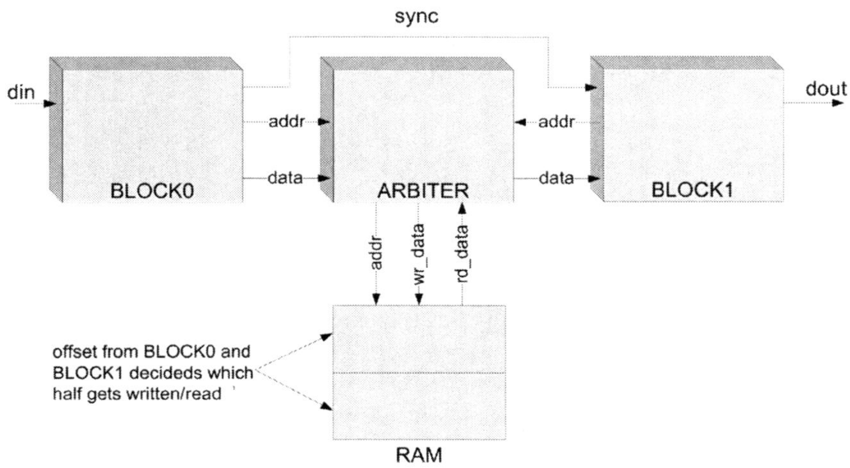

Care must be taken when describing systems like that shown in Figure 9-3. This is because the code typically consists of a combination of both blocking and non-blocking reads.

> **Note**
>
> Reading an empty channel causes a C++ assertion.

The use of non-blocking reads allows us to model parallelism in a sequential C++ description. However sequential calls to functions that communicate via channels can lead to problems because an empty channel cannot be read without asserting during C++ simulation. Changing the order of execution of the functions may sometimes solve this problem, but in general reactive systems cannot be effectively modeled this way. The recommended style is to use a class based approach, where one or more functions are encapsulated in a class. The instance of that class can then be passed to other classes or functions via a reference. Passing a reference to a class allows direct access to that class instance within a function, eliminating the problems associated with sequential execution. Example 9-5 uses this technique for implementing the design shown in Figure 9-3.

Example 9-5. Memory Arbiter

```
1  #include "memclass.h"
2  #include<ac_channel.h>
3  void block0( ac_channel<int> &data_in,
4           ac_channel<bool> &sync,
5           memclass<int,128> &mem){
6    int tmp;
7    static bool buf_sel = false;
8    int offset = buf_sel ? 0:64;
9    for(int i=0;i<64;i++){
10     mem.write_port(i+offset,data_in.read()*10);
11   }
12   sync.write(buf_sel);
13   buf_sel = !buf_sel;
14 }
15 void block1( ac_channel<int> &data_out,
16          ac_channel<bool> &sync,
17          memclass<int,128> &mem){
18   int tmp;
19   bool buf_sel;
20   if(sync.size()>0){
21     buf_sel = sync.read();
22     int offset = buf_sel ? 0:64;
23     for(int i=0;i<64;i++){
24       data_out.write(mem.read_port(i+offset)-1);
25     }
26   }
27 }
28 void top(ac_channel<int> &data_in,
29        ac_channel<int> &data_out){
30   static memclass<int,128> mem;
31   static ac_channel<bool> sync;
32   block0(data_in,sync,mem);
33   block1(data_out,sync,mem);
34 }
```

```
Design Constraints:
Block0 and Block1 mapped to hierarchy
All loops pipelined with II=1
Arbiter member function of memclass mapped to hierarchy.
```

The details of Example 9-5 are:

- Lines 28 and 29 define the top-level function which has an input data channel and an output data channel. These must be defined as references.
- Line 31 - The internal memory and the arbiter are encapsulated inside of a class "memclass". The class is templatized for the data type and number of memory words. This is covered in detail next.

- Line 31 defines the channel used to send the synchronization signal from BLOCK0 to BLOCK1.
- Lines 32 and 33 call the block0 and block1 functions and each function is passed a reference to the memclass variable "mem". By passing a reference to "mem" each function has direct access to the arbiter and internal memory. This prevents reading an empty channel.
- Lines 3 through 5 define the block0 function interface. The arbiter/ memory is passed as a reference on this interface.
- Lines 7 and 8 define a static variable that determines which half of the memory should be written. This variable is used to calculate the offset into the memory.
- Lines 9 though 11 reads the input channel and calls the memclass "write_port" member function. The input data, address, and offset are passed to "write_port".The "write_port" member function accesses the internal memory via the arbiter.
- Line 12 sends the synchronization signal to block1 indicating which half of the memory can be read.
- Lines 15 through 17 define the block1 function interface. This function also has access to the memclass variable via a reference.
- Line 20 tests the "sync" FIFO size to see if the synchronization signal has been sent. NOTE: This has a very important effect on the synthesized hardware for block1. For the C++ simulation this condition is always true since block0 is always executed before block1. However, the synthesized hardware for block1 begins running immediately after reset. Because of this block1 must not be allowed to read until block0 has finished writing the memory. Adding this non-blocking read of "sync" gives the desired behavior.
- Lines 21 and 22 read the "sync" signal when it is available and calculates the offset into the memory.
- Line 25 calls the memclass "read_port" member function with the current address and offset. The "read_port" member function accesses the internal memory via the arbiter.

Example 9-6 shows the memclass class definition.

Example 9-6. Arbiter with Internal Memory Class

```
1   #ifndef __ARBITER__
2   #define __ARBITER__
3   #include <ac_channel.h>
4   #include <ac_int.h>
5   template<typename T, int N>
6     class memclass{
7    private:
8     ac_channel<ac_int<ac::log2_ceil<N>::val,false> > addr_rd;
9     ac_channel<T > data_rd;
10    ac_channel<ac_int<ac::log2_ceil<N>::val,false> > addr_wr;
11    ac_channel<T > data_wr;
12    bool priority;
13    T ram[N];
14   public:
15    memclass():priority(false){};
16      T read_port(ac_int<ac::log2_ceil<N>::val,false> addr);
17     void write_port(ac_int<ac::log2_ceil<N>::val,false> addr,
T data);
18      void arbiter();
19  };
20  #include "memclass_read.h"
21  #include "memclass_write.h"
22  #include "memclass_arbiter.h"
23  #endif
```

The details of Example 9-6 are:

- Line 5 - the class is templatized for data type "T" and number of array elements "N".
- Lines 8 through 11 define channels that are used to connect to both the writer function "block0" and the reader function "block1". Separate channels are needed for each function that needs to access the internal memory.
- Line 12 defines a variable "priority" that is used by the arbiter to decide who gets access to the memory.
- Line 13 defines the internal array "ram" which is mapped to memory and accessed via the arbiter.
- Lines 16 through 18 define the member function prototypes for the class. Only the "arbiter" member function is mapped to hierarchy. The "read_port" and "write_port" methods used by block0 and block1 are left alone and are inlined and synthesized where they are called.

Examples 9-7, 9-8, and 9-9 show the memclass member functions.

Example 9-7. Memclass Read Method

```
1   template<typename T, int N>
2      T memclass<T,N>::read_port(ac_int<ac::log2_ceil<N>::val,false> addr){
3        addr_rd.write(addr);
4        arbiter();
5        return data_rd.read();
6      }
```

The details of Example 9-7 are:

- Line 3 writes the current address "addr" into the memclass's "addr_rd" channel. This channel is read by the arbiter.
- Line 4 calls the arbiter member function.
- Line 5 reads data from the memclass's "data_rd" channel. This channel is written by the arbiter. NOTE: this read is a blocking read and expects the data to be available. If it's possible that the "data_rd" channel may not be written by the arbiter, additional checks must be added. This is covered later.

Example 9-8. Memclass Write Method

```
template<typename T, int N>
    void memclass<T,N>::write_port(ac_int<ac::log2_ceil<N>::val,false> addr, T data){
  addr_wr.write(addr);
  data_wr.write(data);
  arbiter();
}
```

Example 9-8 writes the address and data into the memclass channels that connect to the arbiter. This function should not block as long as the channel FIFOs do not overflow.

Example 9-9. Memclass Arbiter

```
template<typename T, int N>
  void memclass<T,N>::arbiter(){
  bool p[2];
  p[0] = addr_rd.size()!=0;
  p[1] = addr_wr.size()!=0;
```

```
6     if(p[0] & (priority | !p[1])){
7       data_rd.write(ram[addr_rd.read()]);
8       priority = !priority;
9     }
10    else if(p[1]&(!priority | !p[0])){
11      ram[addr_wr.read()] = data_wr.read();
12      priority = !priority;
13    }
14  }
```

The details of Example 9-9 are:

- Lines 4 and 5 check to see if any addresses are available in either the read or write channel and assigned the results to flags p[0] and p[1]. Being able to test whether valid address data is in either channel without actually reading it prevents a read from blocking.
- The arbiter uses a round-robin priority to decided whether the read or write channel is serviced. Each time the reader or writer is serviced the priority is switched.
- Lines 6 and 7 checks to see if the reader has sent an address "p[0]==true" and whether the reader has priority "priority==true" or if the writer is not trying to write "p[1]==false". If the reader is granted access the read channel "data_rd" is written with the data from the internal array "ram" using the address read from the reader "addr_rd.read()".
- Line 8 switches the priority to the writer after each read.
- Lines 10 through 12 check to see if the writer is granted access in the same fashion as the reader. The priority is switched to the reader after each write.

Note

Each call to the arbiter by the read_port function always returns data, so the read never blocks. Thus it is only necessary to call the arbiter once. However, in more complicated designs it is possible that the arbiter does not always return data to the read_port function. In this case more sophisticated coding is required to prevent the read data channel from asserting.

Preventing C++ Assertions from Reading Empty Channels

The previous arbitration example showed how a mixture of both non-blocking reads and blocking reads could be used to model concurrency. The C++ from that example simulates correctly because the arbiter is always able to return data when called by the reader. If this didn't happen the reader would read an empty channel which causes an assertion in C++ simulation. This cannot be allowed to happen.

Consider the following example where the arbiter does not always return data to the reader. This can happen for a number of reasons. Perhaps there is still data in the write channels while the reader is calling the arbiter, or the arbiter may be responding to some other event. In either case the reader must be enhanced. Example 9-10 shows how to code the read_port function so that it never asserts. Now after the address is issued on Line 3, a "while" loop repeatedly calls the arbiter until a non-zero size is detected on the read data channel. The read of the data channel on line 6 does not happen until the arbiter returns valid data.

Example 9-10. Enhanced read_port Member Function

```
1  template<typename T, int N>
2     T memclass<T,N>::read_port(ac_int<ac::log2_ceil<N>::val,false> addr){
3       addr_rd.write(addr);
4       while(!data_rd.size()>0)//while no read data
5         arbiter();
6       return data_rd.read()*37;
7     }
```

Feedback

It was discussed earlier in "Pipeline Feedback" on page 73 that intra-block feedback can prevent the ability to pipeline and schedule a design. With the introduction of channels and hierarchy it is now possible to have inter-block feedback, which can cause the C++ or RTL simulation to assert or deadlock respectively. There is also the possibility that the C++ may simulate while the RTL still deadlocks. This usually happens because feedback in the untimed C++ description is instantaneous, while in the RTL it is dependent on the total latency of the feedback path. A feedback channel will "block" if it is read when empty. Sometimes designs can be made to work using only blocking reads by balancing the feedback latency. However many designs require non-blocking reads to make the system function as desired.

C++ Assertion

The following simple example is used to illustrate how feedback can cause the C++ simulation to assert. Examples 9-11, 9-12, and 9-13 illustrate a simple two block design that contains feedback between the blocks. Although this example is somewhat contrived, it clearly illustrates the problems associated with feedback. The block0 function reads the input data and coefficient from the top-level interface and multiplies them together on line 8 of Example 9-11. It also reads the channel called "feedback" and subtracts that data from the input times the coefficient. However the feedback channel is written on line 11 of the block1 function in Example 9-12. Since block0 is called before block1 in Example 9-13, the feedback channel is empty the first time it is read, and causes an assertion in the C++ simulation. This design never functions as desired.

Example 9-11. Asserting on Empty Feedback Channel Read - BLOCK0

```
1   #include "assert.h"
2   void top(ac_channel<int> &data_in,
3        ac_channel<int> &coeff,
4        ac_channel<int> &data_out){
5     static ac_channel<int> data;
6     static ac_channel<int> feedback;
7     block0(data_in,coeff,data,feedback);
8     block1(data,data_out,feedback);
9   }
```

Example 9-12. Asserting on Empty Feedback Channel Read - BLOCK1

```
1   #include "assert.h"
2   void block1( ac_channel<int> &data_in,
3            ac_channel<int> &data_out,
4            ac_channel<int> &feedback){
5     int fb;
6     int tmp = data_in.read();
7     if(tmp>MAX)
8       fb = tmp - OFFSET;
9     else
10      fb = 0;
11    feedback.write(fb);
12    data_out.write(tmp);
13  }
```

Example 9-13. Asserting on Empty Feedback Channel Read - Top

```
#include "assert.h"
#include<ac_channel.h>
void block0( ac_channel<int> &data_in,
         ac_channel<int> &coeff,
         ac_channel<int> &data_out,
         ac_channel<int> &feedback){
   data_out.write(data_in.read()*coeff.read()-feedback.read());
}
```

Preloading the Channels/FIFOs

Example 9-13 asserts because the empty feedback channel is read the first time block0 executes. The ac_channel class allows channels to be pre-loaded with data using the class constructor.

Usage:

```
ac_channel<int> my_channel(<prefil_number>, <prefill value>)
```

The reason why Example 9-13 asserts is because the first read of the feedback channel is done on an empty channel. Thus if the channel is preloaded with just a single value the first call to block0 can complete without asserting. Example 9-14 shows the code modification of the top-level design. Line 6 declares the "feedback" ac_channel and initializes it to be preloaded with 1 value equal to zero. This allows the C++ to simulate without asserting. However, although the C++ simulates correctly, the RTL deadlocks or stutters. This is discussed in the next section.

Example 9-14. Preloading the Channel

```
1  #include "deadlock.h"
2  void top(ac_channel<int> &data_in,
3       ac_channel<int> &coeff,
4       ac_channel<int> &data_out){
5    static ac_channel<int> data;
6    static ac_channel<int> feedback(1,0);
7    block0(data_in,coeff,data,feedback);
8    block1(data,data_out,feedback);
9  }
```

Deadlock

Example 9-14 solved the problem of asserting when reading an empty channel by allowing the channel to be preloaded with one or more pieces of data. However, there is still a potential for deadlocking or stuttering in the RTL if this is not done correctly, and in fact this is what happens when Example 9-14 is synthesized and simulated in RTL. Figure 9-4 shows the hardware diagram for Example 9-14.

Figure 9-4. Stuttering or Deadlock

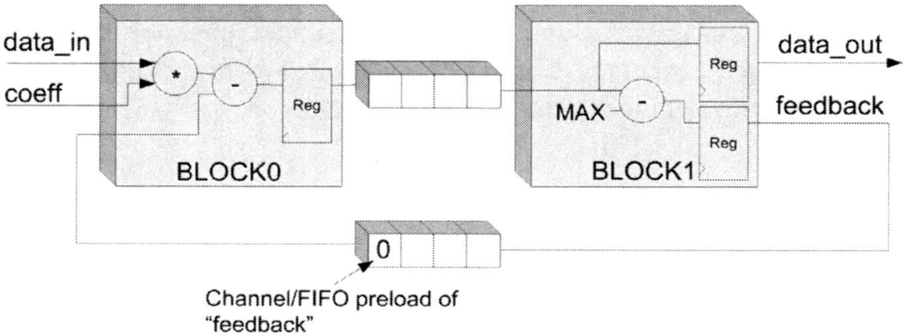

By preloading the feedback FIFO, BLOCK0 can read the FIFO and its inputs and produce its first output. However once the feedback FIFO is read it becomes empty and BLOCK0 must then stall. BLOCK1 can start once BLOCK0 produces an output. BLOCK1 writes the output and the feedback channel and then stalls, waiting for the next value from BLOCK0. BLOCK0 can now start again since the feedback FIFO is no longer empty. Thus in this example BLOCK0 and BLOCK1 "stutter" and data is only produced every other clock cycle. If the latency of either block was greater than one clock cycle the entire system would deadlock.

Note

In general the feedback FIFO should be preloaded with enough data to match the feed forward latency.

Figure 9-5 shows the feedback FIFO preloaded with enough data to keep the design running every clock cycle.

Figure 9-5. Balancing the Feed Forward Latency

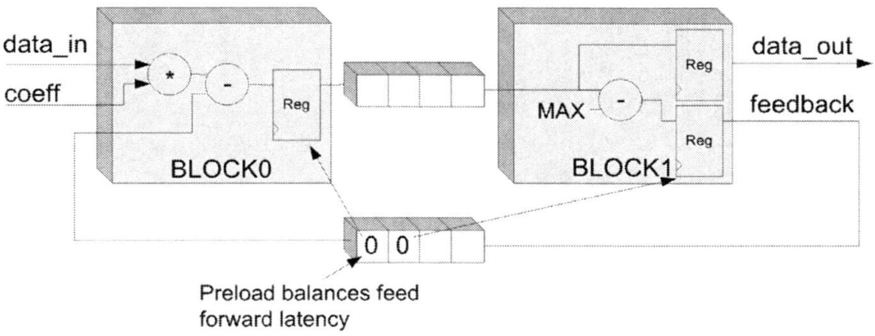

Variable Rate or Data Dependent Feedback

If the feedback rate is variable the system always deadlocks when using blocking reads, and the method of preloading the FIFOs does not help. Example 9-15 shows the block1 function from Example 9-14, but now the feedback channel is only written intermittently. Lines 7 through 10 show that block1 only writes the feedback channel when the data exceeds the value set by MAX. However, block0 still always reads the feedback channel each time it is called. At some point the feedback channel is read when empty (underflow), which causes an assertion or deadlock, illustrating the need for non-blocking reads. 1

Note

If the feedback rate matches the feed forward rate, preloading the FIFOs to match the feed forward latency allows the system to run without deadlocking when using blocking reads. However, if the rates are different, the system always deadlocks when using blocking reads. In this case non-blocking reads must be used.

Example 9-15. Variable Rate Feedback with Blocking Read

```
1   #include "variable.h"
2   void block1( ac_channel<int> &data_in,
3               ac_channel<int> &data_out,
4               ac_channel<int> &feedback){
5     int fb;
6     int tmp = data_in.read();
7     if(tmp>MAX){
8       fb = tmp - OFFSET;
```

```
      feedback.write(fb);
    }
    data_out.write(tmp);
}
```

If non-blocking reads are used, the feedback channel can be read conditionally only when there is valid data in the channel. This eliminates any possibility of asserting or deadlocking due to under flowing the channel/FIFO. Example 9-16 shows the block0 function of Example 9-14 rewritten to use non-blocking reads so that the feedback channel is only read when data is available. Line 6 defines a variable “fb” that is set equal to zero each time the function is called. Lines 7 and 8 check to see if any data is in the feedback channel/FIFO, and reads the data from the channel into “fb” if valid data is present. Otherwise “fb” is left initialized to zero. Line 9 then subtracts “fb” from the multiplication. Thus if there is no data in the feedback channel the function can still complete. One detail to keep in mind is that use of non-blocking reads has the potential to break the one-to-one relationship between the C++ and synthesized RTL. However, this is to be expected since the C++ feedback is available immediately, while the scheduled and synthesized design always has one or more cycles of latency. Other methods of verification can be used to verify the design by testing the RTL output for signal-to-noise ratio, bit error rate, etc.

Example 9-16. Variable Rate Feedback with Non-blocking Reads

```
1   #include "variable.h"
2   void block0( ac_channel<int> &data_in,
3                ac_channel<int> &coeff,
4                ac_channel<int> &data_out,
5                ac_channel<int> &feedback){
6     int fb = 0;
7     if(feedback.size()>0)
8       fb = feedback.read();
9     data_out.write(data_in.read()*coeff.read()-fb);
10  }
```

Chapter 10

Digital Filters

Introduction

Up till this point each chapter has introduced different concepts related to high level synthesis. These concepts included simple C++ examples, both contrived as well as concrete, to illustrate how hardware is synthesized from C++. This chapter builds upon this foundation by covering design concepts using real world designs. There is no better place to begin applying the principles of high level C++ synthesis hardware design than the topic of digital filters. Digital filters are something that most hardware engineers are familiar with, and are ideal for showing off the power of HLS. One of the main reasons for this is that filters tend to have an "explorable" memory architecture, where the use of loop unrolling and pipelining allow designers to tune the area and performance to meet the design specification.

FIR Filters

The finite impulse response (FIR) filter is encountered in a wide range of applications in both communications and video. The theory behind filter design is beyond the scope of this chapter. The intent here is to introduce some common filter structures and concepts, and how to best implement them in C++.

A FIR filter can be expressed as a difference equation:

```
y[n] = h   0x[n] + h1x[n-1] + ... + hNx[n-N]
```

where:

- x[n] is the input signal
- y[n] is the output signal
- hi are the filter coefficients

Another way to look at this is that the output y[n] is a weighted sum of the current and previous values of x[n]. This is often expressed as:

$$y[n] = \sum_{k=0}^{N} h_k x[n-k]$$

The equation above is usually represented in block diagram form, shown below in Figure 10-1 where "i" in this example equals three, meaning that four filter taps are required. The delay elements are realized as a shift register in hardware, and the coefficient multiplies and summation are usually explorable via HLS design constraints.

Figure 10-1. Block Diagram of FIR Filter with i==3

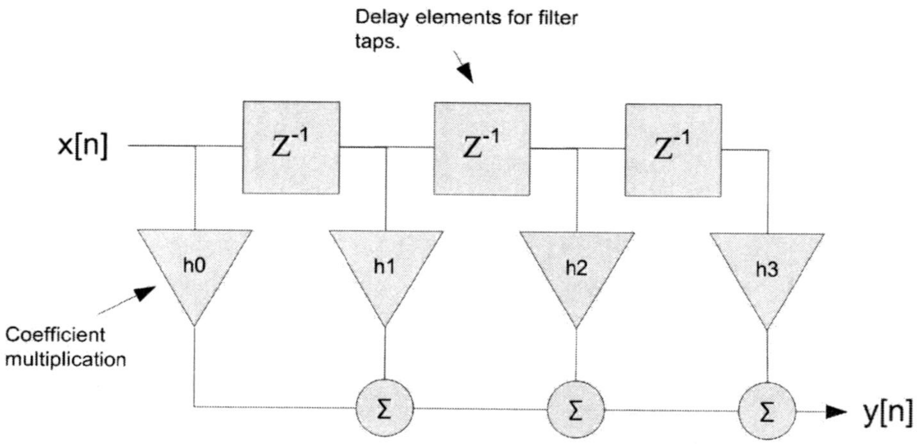

Register Based Filters

Filters that have their delay elements mapped to registers are easy to express in C++. These type of filters have a memory architecture that is highly explorable via loop unrolling and pipelining.

External Coefficients

Example 10-1 shows the C++ implementation for Figure 10-1, where the filter coefficients are read from the top-level design interface.

Example 10-1. FIR Filter with External Coefficients

```
1   #include "fir_filter.h"
2   #include "shift_class.h"
3   void fir_filter (ac_fixed<8,1> *x,
4                    ac_fixed<8,1> h[4],
5                    ac_fixed<19,4> *y){
6     static shift_class<ac_fixed<8,1>,4> regs;
7     ac_fixed<19,4> temp = 0;
8     regs << *x;
9     MAC:for (int i = 0; i<4; i++) {
10      temp += h[i]*regs[i];
11    }
12    *y = temp;
13  }
14

Design constraints:
All IO mapped to wire interfaces with enable
All internal arrays mapped to registers
Main loop pipelined with II=1
Shift register loops fully unrolled
MAC loop left rolled
```

The details of Example 10-1 are:

- Line 2 - the shift register class covered in “Class Based Shift Register” on page 119 is included and used to implement the FIR tap delays.
- Lines 3 through 5 define the input data and coefficients and filter output. Fixed point data types are used for this design.
- Line 6 creates a static instance of the shift register class which is then used as the tap shift register. This is declared static so the tap data persists between function calls. The data type is the same type as the input data.
- Line 8 shifts in new data each time the filter function is called.
- Lines 9 through 11 multiply all taps and coefficients.

The hardware diagram for Example 10-1 is shown below in Figure 10-2. Because the MAC loop is left rolled the multiplier and adder can be shared to compute the filter output. Of course this also means that the filter throughput equals four in this case since each multiply and accumulate takes one clock cycle.

Figure 10-2. FIR Filter with External Coefficients

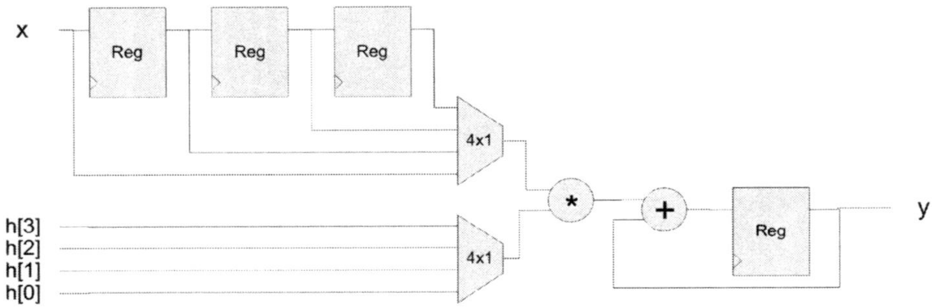

The MAC loop of Example 10-1 can be partially or fully unrolled to increase performance. Figure 10-3 shows Example 10-1 with the loops fully unrolled. It is assumed that the clock is slow enough so that the multipliers and adder tree have been scheduled in the same clock cycle. HLS automatically inserts additional pipeline registers as needed.

Figure 10-3. Fully Parallel FIR Filter with External Coefficients

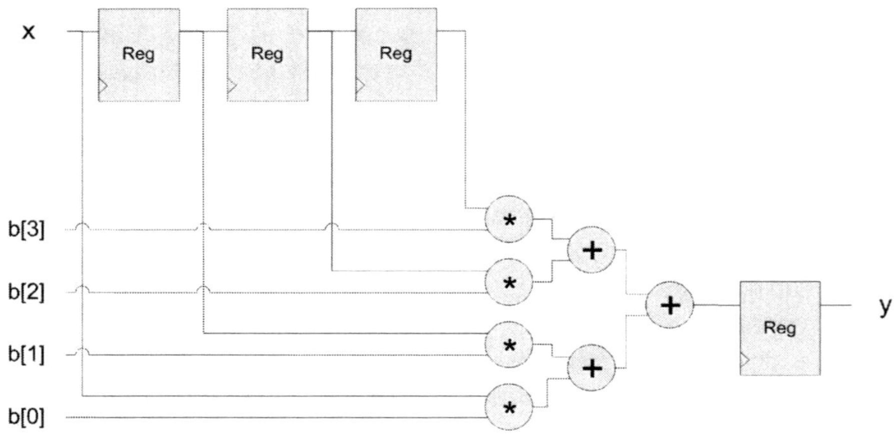

Constant Coefficients

The previous example had external filter coefficients that could be programmed outside of the top-level design. In that example each tap/coefficient multiplication required a hardware multiplier. This can become very costly in area as the size of the filter taps and bit widths increases. In may designs the coefficients do not need to be programmable because they do not change. In other words they are constant. When this is the case the coefficients should be directly coded into the design as constants. This allows HLS to perform constant propagation and optimize the multipliers as constant multipliers. Example 10-2 shows the FIR filter coded with constant coefficients. Line 6 defines a constant array which is initialized with the constant coefficients.

Example 10-2. FIR Filter with Constant Coefficients

```
1   #include “fir_filter.h”
2   #include “shift_class.h”
3   void fir_filter (ac_fixed<8,1> *x,
4                    ac_fixed<19,4> *y){
5       const ac_fixed<8,1> h[4] = {0.30011, 0.90032, 0.90032,
0.30011};
6     static shift_class<ac_fixed<8,1>,4> regs;
7     ac_fixed<19,4> temp = 0;
8     regs << *x;
9    MAC:for (int i = 0; i<4; i++) {
10      temp += h[i]*regs[i];
11    }
12    *y = temp;
13  }
```

Loadable Coefficients

There are two ways to implement filters with loadable coefficients. One is to use an interface synthesis component that contains both storage and a cpu-like interface for programming the coefficients. Another way is to code it directly in the C++ source. This is shown below in Example 10-3. Lines 10 through 12 of this example test the interface variable “ld” to see if the coefficients should be updated from the top-level interface. The loading of the coefficients happens along with the filter computation. Thus the coefficients can be updated and used immediately.

Example 10-3. Filter with Loadable Coefficients

```
1   #include “fir_filter.h”
2   #include “shift_class.h”
3   void fir_filter (ac_fixed<8,1> &x,
4                    ac_fixed<8,1> h[4],
5                    ac_fixed<19,4> &y,
6                    bool &ld){
7     static shift_class<ac_fixed<8,1>,4> regs;
8     ac_fixed<19,4> temp = 0;
9     static ac_fixed<8,1> h_int[4];
10    if(ld==true)
11      for(int i=0;i<4;i++)
```

```
        h_int[i] = h[i];
    regs << x;
    MAC:for (int i = 0; i<4; i++) {
      temp += h_int[i]*regs[i];
    }
    y = temp;
}
```

Symmetric Coefficients

The previous filter examples for both dynamic and constant coefficients were written with no assumptions made about the coefficient properties. In other words the coefficient values could be randomly chosen and the filter would still produce a predictable result. In reality filter coefficients are often related and their properties can be exploited to give better performance and area. Coefficients are often symmetrical around the center tap, or two innermost taps, of the filter. This symmetry allows the number of multiplications to be reduced.

Even Symmetric

Consider the example "FIR Filter with External Coefficients" on page 230 where the coefficients on the interface are:

```
ac_fixed<8,1> h[4] = {0.3, 0.9, 0.9, 0.3};
```

Manually unrolling the MAC loop results in:

```
temp = h[0]*regs[0] + h[1]*regs[1] + h[2]*regs[2] +
h[3]*regs[3];
```

Or:

```
temp = 0.3*regs[0] + 0.9*regs[1] + 0.9*regs[2] + 0.3*regs[3];
```

This can be re-factored to:

```
temp = 0.3*(regs[0] + regs[3]) + 0.9*(regs[1] + regs[2]);
```

By pre-adding the tap values together half the number of multipliers can be eliminated. Example 10-4 shows Example 10-1 rewritten assuming that the coefficients at the interface are symmetrical.

Example 10-4. FIR Filter with Even Symmetry

```
1   #include "fir_filter.h"
2   #include "shift_class.h"
3   void fir_filter (ac_fixed<8,1> *x,
4                    ac_fixed<8,1> h[2],
5                    ac_fixed<19,4> *y){
6     static shift_class<ac_fixed<8,1>,4> regs;
7     ac_fixed<19,4> temp = 0;
8     regs << *x;
9     MAC:for (int i=0;i<4/2;i++){
10      temp += h[i]*(regs[i]+regs[4-1-i]);
11    }
12    *y = temp;
13  }
```

The details of Example 10-4 are:

- Line 4 - because the coefficients are symmetrical only half the coefficient array is needed on the interface.
- Line 9 - the loop only has to run for half the number of iterations as the original loop. This is because the indices into the shift register can be generated simultaneously for the pre-add. So there is an improvement in throughput and latency even if the loop is left rolled.
- Line 10 - the tap registers are symmetrically added together and then multiplied.

Figure 10-4 shows the hardware when the loops are fully unrolled and the main loop is pipelined with II=1.

Figure 10-4. FIR Filter with Even Symmetry

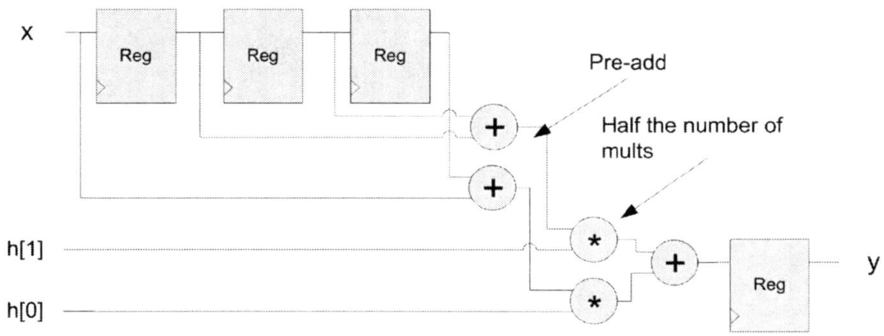

Odd Symmetric

Odd-symmetric filters are coded slightly differently than even symmetric filters. The center tap must be handled separately because there is no pre-add. Other than that the approach is the same as covered above. For a 5-tap filter assume the following coefficients, which are symmetrical about the center coefficient.

```
ac_fixed<8,1> h[5] = {0.078, 0.253, 0.335, 0.253, 0.078};
```

Manually unrolling the MAC loop results in:

```
temp = h[0]*regs[0]+h[1]*regs[1]+h[2]*regs[2]+h[3]*regs[3]+h[4]
*regs[4];
```

This can be re-factored to:

```
temp = 0.78*(regs[0] + regs[4])+0.335*regs[2]+0.253*(regs[1] +
regs[3]);
```

Example 10-5 shows an example of a 5-tap FIR filter where the coefficients are symmetric. Lines 11 through 14 handle the pre-add of the taps. The center tap is simply passed through, which can be seen in the hardware diagram of Figure 10-5.

Example 10-5. FIR Filter with Odd Symmetry

```
1  #include "fir_filter.h"
2  #include "shift_class.h"
3  void fir_filter (ac_fixed<8,1> *x,
4                   ac_fixed<8,1> h[3],
5                   ac_fixed<19,4> *y){
6    static shift_class<ac_fixed<8,1>,5> regs;
7    ac_fixed<19,4> temp = 0;
8    ac_fixed<9,2> sum = 0;
9    regs << *x;
10   MAC:for (int i=0;i<5/2+1;i++){
11     if(i==2)
12       sum = regs[5/2];//middle tap
13     else
14       sum = regs[i]+regs[5-1-i];
15     temp += h[i]*sum;
16   }
17   *y = temp;
18 }
```

Figure 10-5. FIR Filter with Odd Symmetry

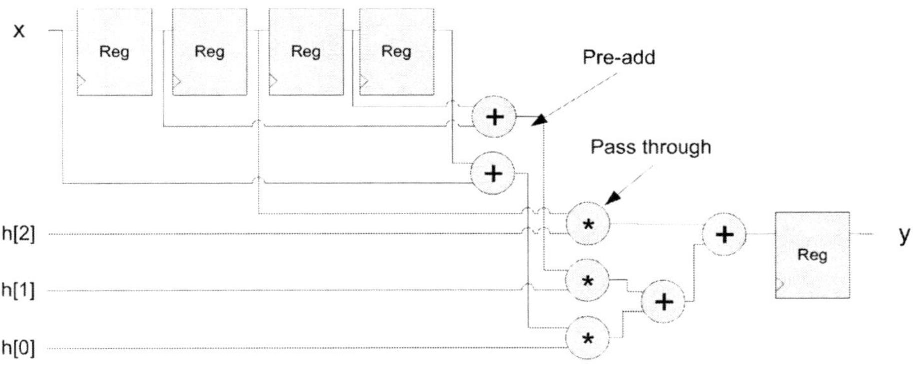

Transposed

The previous discussion of FIR filters was based on the direct form FIR, where the tap-values are multiplied against the coefficients and accumulated. For the fully parallel implementations this results in an adder tree, shown in Figure 10-3. Some of the disadvantages of an adder tree are that it can have multi-cycle latency for high clock speeds and large number of taps. It may also be more difficult to route due to the large number of interconnects between the adders. An alternative implementation is the transposed form of the FIR, which has single cycle latency independent of the number of taps. However it is more limited in terms of FMax due to fanout. Figure 10-6 presents a DSP block diagram that shows the general structure and data flow of a transposed FIR. The figure shows that rather than shifting the input data, the partial products of the current input "x" and the coefficients is shifted and accumulated.

Figure 10-6. Block Diagram of Transposed FIR

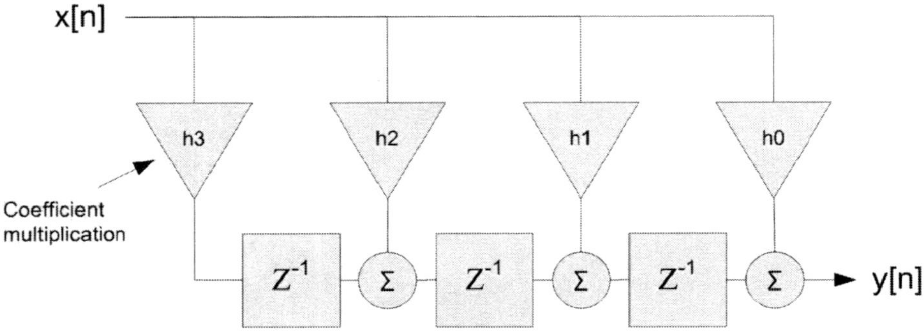

Example 10-6 shows the transposed FIR implementation of a four tap filter.

Example 10-6. Transposed FIR Filter

```
1   #include "fir_filter.h"
2   void fir_filter (ac_fixed<8,1> *x,
3                    ac_fixed<8,1> h[4],
4                    ac_fixed<19,4> *y){
5     static  ac_fixed<19,4> regs[4];
6     ac_fixed<19,4> temp = 0;
7     MAC:for (int i=3; i>=0; i--) {
8       if(i==0)
9         temp = 0;
10      else
11        temp = regs[i-1];
12      regs[i] = *x * h[3-i] + temp;
13    }
14    *y = regs[3];
15  }

Design constraints:
All IO mapped to wire interfaces
All internal arrays mapped to registers
Main loop pipelined with II=1
MAC loop fully unrolled
```

The details of Example 10-6 are:

- Line 5 - the shift register class can't be used in this design so a static array is declared internally. The bit width is set to account for the bit growth due to the multiply and accumulation. In general the transposed FIR has larger register area than the direct form which only needs to be as wide as the input data.

- Lines 7 through 13 - Starting with the right most register, the partial products are computed, accumulated, and then stored.

Figure 10-7 shows the synthesized hardware for Example 10-6. The synthesized hardware illustrates both the benefits and drawbacks of the transposed implementation. The benefit is that the latency is always equal to one regardless of the clock frequency. Each read of a new input "x" produces an output on the next clock edge. However, there is a limit on how fast you can clock this implementation because "x" fans out to ever tap.

Figure 10-7. Hardware for Transposed FIR

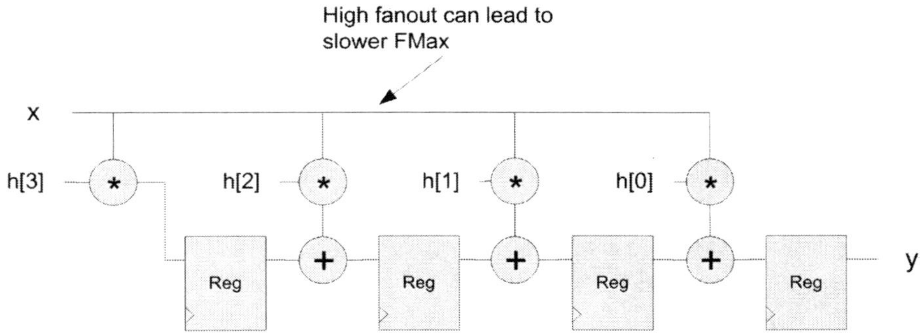

Systolic

It was shown that the direct form and transposed FIR implementations both have benefits as well as limitations. The direct form can run at higher clock frequencies, but at the cost of longer latency as well as more complex routing. The transposed always has a latency of one but is limited in FMax based on the total number of taps. The third option is to use a systolic array implementation, which should have the fastest clock frequency and routing, but longer latencies. The systolic architecture is presented here without explanation since this topic is well beyond the scope of this book. What is important is to understand the impact of using this architecture as well as how to code it in C++. Figure 10-8 shows the hardware diagram of a systolic implementation of a four-tap FIR filter. The implementations consists of an array of processing elements (PE) that are cascaded. Inspection of the figure shows that the worst-case timing path is limited to just the multiply and adder for each PE. However it can also be seen that the latency is proportional to the number of taps.

Figure 10-8. Systolic Architecture

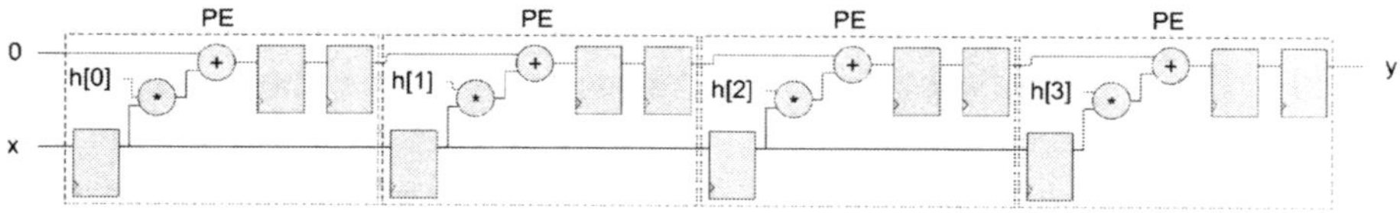

A C++ class based approach can be used to allow efficient implementation of the systolic architecture shown in Figure 10-8. Since each PE is identical it would seem that creating a PE class and then instantiating it multiple times would be the most efficient solution. Example 10-7 shows the C++ class that implements one of the PEs shown in Figure 10-8. An important thing to point out with this example is that describing behavior at a fairly low level is still possible using C++. The number of register stages is explicitly coded into the class. However all this low level detail can be abstracted away when the class is used to construct the filter. Thus the low level details are written once to describe the PE, after which they become transparent to the design.

Example 10-7. Systolic Processing Element Class

```
1   template<typename T0, typename T1, typename T2>
2   class pe_class{
3   private:
4     T0 x;
5     T2 y0;
6     T2 y1;
7   public:
8     void exec(T0 &x_in, T1 &h, T2 &y_in,T0 &x_out, T2 &y){
9       y = y1;
10      x_out = x;
11      y1 = y0;
12      y0 = x * h + y_in;
13      x = x_in;
14    }
15  };
```

The details of Example 10-7 are:

- Line 1 - the class is templatized for the input, coefficient, and output data types.

- Line 4 through 6 define the member variables that implement the three shift registers in the PE of Figure 10-8.
- Lines 9 and 10 - the registered values of x and y are written to the PE output.
- Lines 11 and 12 - the two output registers on “y” are implemented. The PE mult-add feeds the first register stage of “y”.
- line 13 - the current value of “x_in” is stored in a register.

Example 10-8 shows how the systolic FIR is implemented using the PE class.

Example 10-8. Systolic FIR implementation

```
1   #include "pe_class.hpp"
2   #include "fir_filter.h"
3   void fir_filter (ac_fixed<8,1> *x,
4                    ac_fixed<8,1> h[4],
5                    ac_fixed<19,4> *y){
6      static pe_class<ac_fixed<8,1>,ac_fixed<8,1>,
      ac_fixed<19,4> > pe[4];
7     ac_fixed<8,1> x_int[4];
8     ac_fixed<19,4> y_int[4];
9     ac_fixed<19,4> tmp = 0;
10
11   CONN:for(int i=0;i<4;i++)
12      if(i==0)
13        pe[0].exec(*x, h[i],tmp,x_int[i],y_int[i]);
14      else
15           pe[i].exec(x_int[i-1], h[i],y_int[i-1],
        x_int[i],y_int[i]);
16    *y = y_int[3];
17  }

Design constraints:
All IO mapped to wire interfaces
All internal arrays mapped to registers
Main loop pipelined with II=1
CONN loop fully unrolled
```

The details of Example 10-8 are:

- Line 1 - the PE class from Example 10-7 is included.
- Line 6 - a static four element array of the “pe_class” is defined. The pe_class template arguments are set based on the input, coefficient, and output data types. The array is defined as static because the internal registers in the PEs must persist between function calls to “fir_filter”.
- Lines 7 and 8 define temporary arrays to connect the PEs.

- Lines 11 through 15 - A loop iterates from zero to four (the number of PEs) and connects the PEs together. The first iteration "i==0" feeds the top-level input "x" to the first PE and sets the "y" input equal to zero for the first PE.

Multi-rate Filtering

Multi-rate filters use different sample rates within a system to achieve more area efficient designs. The general principle is to convert a signal to a lower sample rate (down sample or down convert), process at the lower rate, and then convert back to the original rate (up sample or up convert). Processing the signal at a lower rate means fewer multipliers and adders are required to implement the algorithm.

Decimation

Decimation, or down sampling, is used to lower the sample rate of a signal. This is done by periodically discarding enough samples to match the desired rate reduction. A reduction by M means keeping every Mth sample and throwing away the rest. However simply throwing the data away is not usually possible due to frequency aliasing. As the sample rate is reduced the replicated frequency spectra of the sampled signal comes together and overlaps at some point, making the signal unusable. To keep this from happening the signal must be lowpass filtered to prevent aliasing. By combining the lowpass FIR filter with the discarding of samples, the FIR can be made to operate at the lower data rate, reducing area.

There are two approaches towards designing a decimator in C++. One is to use algorithmic code along with loop pipelining to get the desired down sampled rate, the other is to manually code resource sharing into the C++ based on the down sampled data rate.

Algorithmic Decimation

The key to designing an efficient decimation filter when coding more abstractly is to make sure that you pipeline the design at the down sampled rate while reading the data at the original rate. This allows scheduling to share operations across multiple clock cycles, minimizing the area. Examples 10-9 shows a templatized implementation of a decimation filter.

Readers should be familiar with templatization at this point, which allow construction of highly reusable designs. The decimation FIR covered here is used in later sections as well.

Example 10-9. Templatized Decimation FIR

```
1   #include <ac_fixed.h>
2   #include <ac_channel.h>
3   #include "shift_class.h"
4   template<int W0, int W1, int N>
5   struct _WN{
6     enum { val = W0 + W1 + ac::log2_ceil<N>::val };
7   };
8   template<int ID,
9        int W0, int I0,
10       int W1, int I1,
11       int N, int RATE>
12  void dec(ac_channel<ac_fixed<W0,I0> > &x,
13       ac_fixed<W1,I1> h[N],
14        ac_channel<ac_fixed<_WN<W0,W1,N>::val,
        _WN<I0,I1,N>::val> > &y){
15    static shift_class<ac_fixed<W0,I0>,N> regs;
16    ac_fixed<_WN<W0,W1,N>::val,_WN<I0,I1,N>::val> acc = 0;
17    ac_fixed<W0,I0> x_int;
18    READ:for(int i=0;i<RATE;i++)
19      x_int = x.read();
20      regs << x_int;
21    MAC:for (int i = 0; i<N; i++) {
22      acc += h[i]*regs[i];
23    }
24    y.write(acc);
25  }
```

The details of Example 10-9 are:

- Lines 4 through 7- a helper struct is created to calculate the number of integer and fractional bits for the output and internal accumulator data types. The assumption is that the required number of bits equals the sum of the input bits plus the log2 bit growth based on the number of taps N.
- Lines 8 through 11 - the function template parameters allow specification of the input data and coefficient fixed points widths and signedness, the number of taps N, and the decimation rate RATE. The ID parameter is needed for creating multiple unique instances of the function.
- Lines 14 and 16 - the helper struct _WN is used to calculate the required number bits for the output and internal accumulator.

- Lines 18 and 19 - the input data should be read in the beginning of the function call. It is read RATE times where RATE is the decimation rate. The READ loop should be fully unrolled. If the top-level design is pipelined with II=RATE, the RATE reads of the input are spread out during scheduling allowing resources to be shared.
- Line 24 - the write of the output happens every RATE clock cycles if the design is pipelined with II=RATE.

Example 10-10 shows the templatized decimation filter function instantiated in a top-level design. The template parameters are set to eight bit data and coefficients and four filter taps. The decimation rate is set equal to two.

Example 10-10. Using the Templatized Decimation Filter

```
#include <ac_channel.h>
#include "shift_class.h"
#include "fir_filter.h"
#include "decimate.hpp"
void fir_filter(ac_channel<ac_fixed<8,1> > &x,
                ac_fixed<8,1> h[4],
                ac_channel<ac_fixed<18,4> > &y){
  dec<0,8,1,8,1,4,2>(x,h,y);
}

Design constraints:
All IO mapped to wire enable interfaces
IO input rate=1 sample/clock
All arrays mapped to registers
Main loop pipelined with II=2
All loops fully unrolled
```

Figure 10-9 shows the approximate schedule for Example 10-10 and the constraints listed above. The schedule shows that the multiplications of the four filter taps can be distributed over the schedule based on the II, requiring only two multipliers in this case. For this design example the II was set equal to two because the filter decimates by two and the input data rate equals one. In general the II should be set equal to the output data rate which can be calculated as:

```
Output Rate = Input Rate * Decimation Rate
```

Figure 10-9. Schedule of Decimation by Two

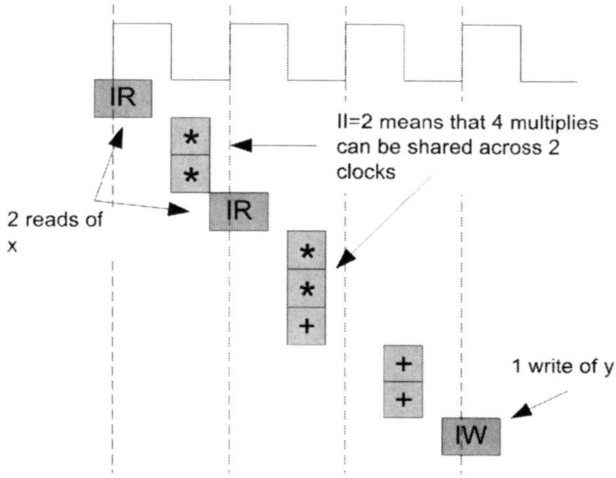

Note

Reading multiple inputs and pipelining at the output rate is only intended for single block decimation. If the design contains multiple decimation filters within the same block of hierarchy a different approach is needed, and is covered in the upcoming sections.

Manual Decimation

The previous example showed how pipelining can be used to synthesize a decimator from a straightforward FIR description. HLS automatically is able to share resources when pipelining with II greater than one. It is also possible to directly code resource sharing into a decimating filter. While it requires more effort, it may have some benefits in terms of smaller area when designing multi-stage decimation with a single block. However, unlike the high-level example which requires almost no understanding about the mechanics of decimation, the low-level implementation requires that we understand how data is processed. The key is to understand how data moves through the tap shift register, and when output data is produced by the filter. This is best understood by looking at the pure algorithmic implementation, shown in Example 10-11. This example would have to be pipelined at the input rate since only one value is read per function call. The output is written conditionally every other call. This may be an inefficient implementation since the full MAC computation is required when “cnt” equals one, and it is pipelined at the input rate. When “cnt” equals zero the data is just shifted.

Example 10-11. Pure Algorithmic Decimation

```
#include <ac_channel.h>
#include "shift_class.h"
void dec2(ac_channel<ac_fixed<8,1> > &x,
                ac_fixed<8,1> h[4],
                ac_channel<ac_fixed<19,4> > &y){
  static shift_class<ac_fixed<8,1>,4> regs;
  ac_fixed<19,8> temp = 0;
  static ac_int<1,0> cnt;

  regs << x.read();
  MAC:for (int i = 0; i<4; i++) {
    temp += h[i]*regs[i];
  }
  if(cnt==1)//Phase 1
    y.write(temp);
  cnt++;
}
```

Figure 10-10 shows the contents of the tap shift register for two calls to the function in Example 10-11. Decimating by M creates M phases, where the output is discarded every M-1 phases, and one phase where the output is computed and written out. Since the discard phase only shifts data it would seem like a good idea to distribute the computation across all phases, reducing the total number of required multipliers and adders. This is what scheduling can do automatically when pipelining with II greater than one.

Figure 10-10. Understanding Decimation

The filter computation of y[n] in Phase 1 is:

```
Phase1: y[5] = h[0]*regs[0] + h[1]*regs[1] + h[2]*regs[2] +
h[3]*regs[3]
```

This can be rewritten to compute the multiply and accumulate of the last two taps in Phase 0. The tap data is read from the tap shift register accounting for the difference in position between the phases.

```
Phase 0: temp = h[2]*regs[1] + h[3]*regs[2]
```

Thus:

```
Phase1: y[5] = h[0]*regs[0] + h[1]*regs[1] + temp
```

Example 10-12 shows a four-tap decimation filter that manually codes sharing across clock cycles. The input data rate for this example is assumed to be one sample per clock. Unlike Example 10-9, this example is pipelined with II=1 and conditionally writes the output based on an internal count. The sharing is coded into the design by manually computing part of the filter during each phase.

Example 10-12. Manual Decimation by Two, Input Rate = 1 samp/clock

```
1   #include <ac_channel.h>
2   #include "shift_class.h"
3   #include "fir_filter.h"
4   void dec_i1(ac_channel<ac_fixed<8,1> > &x,
5                      ac_fixed<8,1> h[4],
6                      ac_channel<ac_fixed<19,4> > &y){
7     static shift_class<ac_fixed<8,1>,4> regs;
8     static ac_fixed<19,8> acc;
9     static ac_int<1,false> cnt;
10
11    regs << x.read();
12    MAC0:for (int i = 0; i<2; i++) {
13      acc += h[i+((1-cnt)<<1)]*regs[i+1-cnt];
14    }
15    if(cnt==1){
16      y.write(acc);
17      acc = 0;
18    }
19    cnt++;
20  }

Design constraints:
All IO mapped to wire enable interfaces
All arrays mapped to registers
Main loop pipelined with II=input rate
MAC loop fully unrolled
```

The details of Example 10-12 are:

- Line 11 - the input data is read once per function call and the design is pipelined with II=input rate which means the synthesized hardware matches the upstream data rate.

- Lines 12 through 14 - since this filter decimates by two the computation of all four taps can be divided into two parts. A one bit counter "cnt" is used to offset the indices into the tap shift registers and coefficient array. The static variable "temp" accumulates the result calculated in each phase.
- Lines 15 through 17 - the output is conditionally written, in other words samples are discarded, based on the decimation rate. Once the output is written the accumulator is cleared and the process starts over.

Example 10-12 showed how to explicitly share resources based on the input and output rate. This decimate by two design is hard coded for an input rate of 1 sample/clock or II=1. If the input rate is not 1 sample/clock, but the design must still be pipelined with II=1, then the design must be further refined to explicitly share resources based on the ratio of input to output rate. The reasons for coding in this fashion become obvious when performing multi-stage decimation within a single block design. Example 10-13 shows a manual approach to a four tap decimation filter class that decimates by two, but has an input rate of two. This class is reused when talking about multi-stage decimation in a later section.

Example 10-13. Decimation Class for Decimate by 2, Input Rate = 2

```
#ifndef _DEC2_H
#define _DEC2_H
#include <ac_fixed.h>
#include "shift_class.h"
template<int W0, int I0, int W1, int I1>
class dec2_i2{
private:
  shift_class<ac_fixed<W0,I0>,4> regs;
  ac_fixed<W0+W1,I0+I1+2> acc;
  ac_int<2,false> cnt;
  bool vld;
  bool go;
public:
  dec2_i2():vld(false), acc(0), go(false), cnt(0){}
  bool exec(ac_fixed<W0,I0> &x,
            ac_fixed<W1,I1> h[4],
            ac_fixed<W0+W1+2,I0+I1+2> &y,
            bool &vld_in,
            bool &vld_out){
    vld = false;
    if(vld_in)
      go = true;
    if(go){
```

```
24        if(!(cnt&1))//read with rate 2
25          regs << x;
26        acc += h[cnt + 2 - (cnt[1]<<2)]*regs[cnt+(1>>cnt[1])-
cnt[1]];
27        if(cnt==3){//write with rate 4
28          y = acc;
29          acc = 0;
30          vld = true;
31        }
32        cnt++;
33        vld_out = vld;
34      }
35    }
36  };
37  #endif
```

The details of Example 10-13 are:

- Line 5 - the class is templatized to allow the integer and fractional bit widths of the input data and coefficients to be specified. The class adds the appropriate bits to account for internal bit growth.
- Line 10 - the "cnt" data member keeps track of which filter phase is being computed. Since the input rate equals two and the decimation rate equal two, four phases are required.
- Lines 15 through 19 - the "exec" function is the class member function that performs the filtering. The interface variables are defined in terms of the class template parameters, and the full precision is maintained at the output. The "vld_in" and "vld_out" variables are used to synchronize the flow of data into and out of the filter.
- Lines 21 through 23 - The filter is implemented such that it does nothing until the first valid data is detected. Once this happens the "go" bit is set true and the filter runs forever.
- Lines 24 and 25 - the input data is read every other call to the function, or when "cnt" equals zero or two.
- Line 26 - four phases means that only one multiplier is required to implement the four tap filter. The "cnt" variable is used to compute the index into the tap shift register and coefficients similar to what was shown in Figure 10-10 on page 244.
- Lines 27 through 31 - every fourth call to the function "cnt==3" the output is written and the "vld_out" flag is set equal to true.

Interpolation

Interpolation, or up sampling, is used to increase the sample rate of a signal. This is accomplished by inserting one or more zeros between each sample. Up sampling by a factor of L means inserting L-1 zeros between each sample. Similar to the decimation filter, interpolation filters have the potential of distortion due to replication of the original frequency spectrum into frequencies in the interpolated spectrum. To prevent this the original signal must be low pass filtered. The low pass filtering can be performed at the input rate, reducing computational overhead.

Algorithmic Interpolation

The construction of a templatized interpolation filter is very similar to the decimation filter, shown below in Example 10-14.

Example 10-14. Interpolation Filter

```
#include <ac_channel.h>
#include "shift_class.h"
template<int W0, int W1, int N>
struct _WN{
  enum { val = W0 + W1 + ac::log2_ceil<N>::val };
};
template<int ID,
     int W0, int I0,
     int W1, int I1,
     int N, int RATE>
void inter(ac_channel<ac_fixed<W0,I0> > &x,
       ac_fixed<W1,I1> h[N],
          ac_channel<ac_fixed<_WN<W0,W1,N>::val,
       _WN<I0,I1,N>::val> > &y){
  static shift_class<ac_fixed<W0,I0>,N> regs;
  ac_fixed<_WN<W0,W1,N>::val,_WN<I0,I1,N>::val> acc = 0;

  WRITE:for(int i=0;i<RATE;i++){
    if(i==0)
      regs << x.read();
    else
      regs << 0;
  MAC:for (int j = 0; j<N; j++) {
      acc += h[j]*regs[j];
    }
    y.write(acc);
    acc = 0;
  }
}
```

The details of Example 10-14 are:

- Lines 3 through 15 are identical to the description of Example 10-9 on page 241.
- Line 17 - the WRITE loop controls the interpolation rate, and iterates RATE times. Each iteration of this loop produces a new output. This loop should be completely unrolled.
- Lines 18 though 21 - each call to the interpolate function only reads one input, which happens in the WRITE loop when i==0. This input is shifted into the tap shift register. All other iterations of the write loop shift zeros into the tap shift register.
- Lines 22through 24 - The MAC loop is completely unrolled and pipelining II is set to match the input rate, allowing scheduling to share resources. This is very similar to the approach that was taken with decimation except that decimation pipelined based on the output rate. Constant propagation occurs on the zeros in the tap shift register because both the WRITE and MAC loops are unrolled. This minimizes the required number of multipliers.
- Line 25 - an output is produced for each iteration of the WRITE loop. A total of RATE outputs are produced for each function call.

Example 10-15 shows the templatized interpolation filter function instantiated in a top-level design. The template parameters are set to eight bit data and coefficients and four filter taps. The interpolation rate is set equal to two.

Example 10-15. Using the Templatized interpolation Filter

```
#include "shift_class.h"
#include "fir_filter.h"
#include "interpolate.hpp"
void fir_filter(ac_channel<ac_fixed<8,1> > &x,
                ac_fixed<8,1> h[4],
                ac_channel<ac_fixed<18,4> > &y){

  inter<0,8,1,8,1,4,2>(x,h,y);
}
```

```
Design constraints:
All IO mapped to wire enable interfaces
All arrays mapped to registers
Main loop pipelined with II=input rate, input rate = 2
All loops fully unrolled
```

Figure 10-11 shows the approximate schedule for Example 10-15. Pipelining at the input rate, II=2, allows resources to be shared during scheduling. In addition to that unrolling both the WRITE loop and the MAC loop allows half the number of multipliers to be optimized away since they are multiplying by zero.

Figure 10-11. Schedule of Interpolation by Two

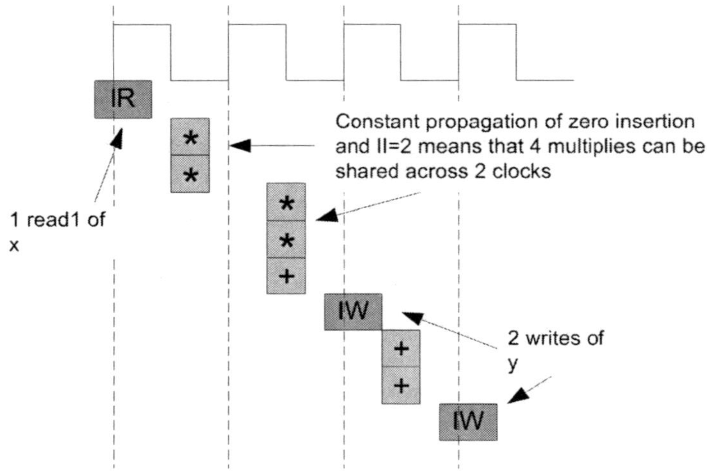

Manual Interpolation

In the same fashion as decimation, interpolation filters can also be written using a more manual approach. This approach also requires analyzing the data movement through the tap shift register to manually code sharing for the most area efficient implementation. Figure 10-12 shows the data movement through the tap shift register for interpolation by two. The main point illustrated by this figure is that half the tap values are zero for an interpolate by two filter. Each filter phase can take advantage of this when computing its output.

Figure 10-12. Manual Interpolation

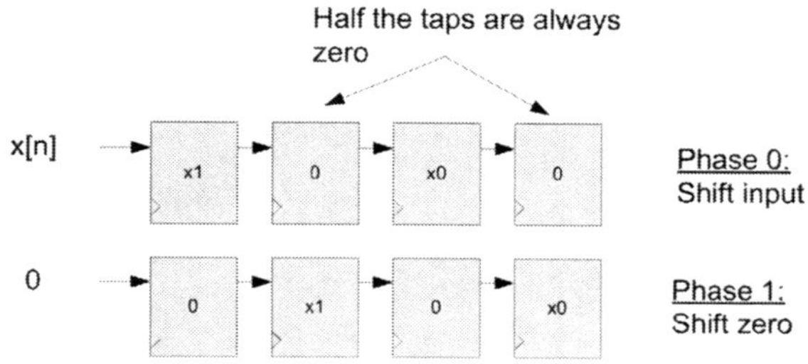

Writing output the filter equations for Figure 10-12 results in:

Phase 0:

```
y[2] = h[0]*x[1] + h[1]*0 + h[2]*x[0] + h[3]*0 = h[0]*x[1] +
h[2]*x[0]
```

Phase 1:

```
y[3] = h[0]*0 + h[1]*x[1] + h[2]*0 + h[3]*x[0] = h[1]*x[1] +
h[3]*x[0]
```

In other words each filter phase only needs two multipliers to compute the output. In general the total number of multipliers is proportional to TAPS/L, where TAPS is the number of filter taps and L is the interpolation factor. Example 10-16 shows the C++ implementation of an interpolate by two filter using manual coding methods. Although this is written at a much lower level than the previous version it has the advantage of explicitly coding sharing which may provided slightly better area.

Example 10-16. Manual Interpolation

```
#include <ac_channel.h>
#include "shift_class.h"
#include "fir_filter.h"
#include "shift_class.h"
void fir_filter (ac_channel<ac_fixed<8,1> > &x,
                 ac_fixed<8,1> h[4],
                 ac_channel<ac_fixed<19,4> > &y){
  static shift_class<ac_fixed<8,1>,4> regs;
  static ac_fixed<19,8> temp;
  static ac_int<1,false> cnt;
  if(cnt==0)
    regs << x.read();
  else
    regs << 0;
```

```
15    MAC0:for (int i = 0; i<2; i++) {
16      temp += h[i*2+cnt]*regs[i*2+cnt];
17    }
18    y.write(temp);
19    temp = 0;
20    cnt++;
21  }
```

```
Design constraints:
All IO mapped to wire enable interfaces
All arrays mapped to registers
Main loop pipelined with II=input rate
MAC loop fully unrolled
```

The details for Example 10-16 are:

- Line 10 defines a 1-bit counter that is used to control which filter phase is active.
- Lines 11 through 14 - read and shift input data on the first phase and insert and shift zeros on all other phases.
- Lines 15 through 17 - only compute non-zero taps for each phase. The phase count "cnt" is used to compute the offset into the tap shift register.
- Line 18 - each call to the filter produces an interpolated output.
- Line 20 - each call to the filter advances the phase count.

Multi-stage Decimation

The previous section showed several approaches towards performing decimation in a single function/block. However it is usually the case that several decimation filters are cascaded together to achieve higher decimation rates. One of the challenges of doing this is coding in such a way as to maximize the amount of sharing that can happen. If a decimation filter has an input rate of one sample/clock its resources cannot be shared with other filters because a filter phase is computed every input clock. However when the input rate is greater than one, resources can often be shared. The slower the input rate and the higher the total decimation rate directly influences how much resources can be shared. The actual amount of sharing depends on the coding style used to implement the cascaded decimation filters.

Multi-block

If sharing of resources between cascaded decimation filters is not needed, the simplest implementation is to use explicit hierarchy for each filter. The filter implementation can then be more algorithmic, like what was shown in Example 10-9 on page 241. Consider the following two stage decimation example, where each stage decimates by two. The top-level input rate is one sample per clock. This example uses the algorithmic decimate by two example that was covered in Example 10-9.

Example 10-17. Multi-Block, Multi-stage Decimation

```
1   #include "decimate.hpp"
2   void dec2_2stage(ac_channel<ac_fixed<8,1> >  &x,
3                    ac_fixed<8,1> h[4],
4                    ac_channel<ac_fixed<28,7> > &y){
5     static ac_channel<ac_fixed<18,4> > y_int;
6
7     BLOCK0:dec<0,8,1,8,1,4,2>(x,h,y_int);
8     if(y_int.available(2))
9       BLOCK1:dec<1,18,4,8,1,4,2>(y_int,h,y);
10  }

Design constraints:
All IO mapped to wire enable interfaces
All arrays mapped to registers
Both instances of "dec", BLOCK0 and BLOCK1, mapped to hierarchy
All loops fully unrolled
Block0 pipelined with II=2
BLOCK1 pipelined with II=4
```

In Example 10-17 BLOCK0 is pipelined with the output rate, or II=2. This is because the input to BLOCK0 comes every clock, and the output every other clock. This requires two multipliers to implement the BLOCK0 filter, similar to what was shown in Figure 10-9 on page 243. However, BLOCK1 now has an input rate of every other clock. Since it also decimates by two, the output rate equals four. This is why BLOCK1 is pipelined with II=4, which in turn means that only one multiplier is required since there are four clock cycles between each output. Lastly, line 8 of Example 10-17 checks the number of elements in the channel FIFO and only calls the BLOCK1 function when there are two elements in the FIFO. This is done for C++ simulation purposes only to prevent assertions caused by reading an empty channel. The hardware is synthesized with a handshake that causes BLOCK1 to stall until data is ready to be read. The use of "available" is required in this example since we want BLOCK1 to begin running as soon as data is available. The "size" method would not work in this case since BLOCK1 would have to wait until there were two elements in the channel FIFO, causing the design to stutter.

Note

The use of the ac_channel available member function is to prevent assertions in the C++ simulation. It is synthesized as a handshake in hardware so care must be taken when using available to prevent deadlock.

Note

Reducing the input rate of a block of hierarchy allows pipelining with an II greater than one, allowing resources to be shared.

Single-block

Combining multiple decimation filters within a single block can increase resource sharing by taking advantage of the decreasing rate of computation for each stage. However, the amount of resource sharing is related to both the overall decimation rate and the style in which the cascade of filters is described.

The most straightforward approach is to use an algorithmic coding style, along with manually generating a rate count in order to make the execution of the different filters mutually exclusive. Mutual exclusivity allows the resources of the filters to be shared. If the first filter stage runs with an input rate of one sample/clock it is not shared, and can be excluded from the following examples. Because of this, these examples assume that the input rate into the block is one sample every four clocks so that the concept of coding for sharing can be better illustrated. Example 10-18 shows a templatized decimate by two function that is used to build a multi-stage decimator. This decimator design was covered previously in Example 10-11 on page 244.

Example 10-18. Templatized Decimate by Two

```
#include <ac_channel.h>
#include <ac_fixed.h>
#include "shift_class.h"
template<int ID, int W0, int I0, int W1, int I1>
void dec2(ac_fixed<W0,I0> &x,
          ac_fixed<W1,I1> h[4],
```

```
             ac_fixed<W0+W1+2,I0+I1+2> &y){
  static  shift_class<ac_fixed<W0,I0>,4> regs;
  ac_fixed<W0+W1,I0+I1+2> acc = 0;
  static ac_int<1,0> cnt;
  regs << x;
 MAC:for (int i = 3; i>=0; i--) {
    acc += h[i]*regs[i];
  }
  if(cnt==1)//Phase 1
    y = acc;
  cnt++;
}
```

Example 10-19 shows a two-stage, single block, decimator design. Each stage decimates by two. The input data rate is assumed to be one input every four clock cycles.

Example 10-19. Single-Block High-Level Decimation

```
#include "dec2_alg.h"
void dec2_2stage(ac_fixed<8,1>   &x,
                 ac_fixed<8,1> h[4],
                 ac_fixed<28,7>  &y){
  static  ac_fixed<18,4>  y0_int=0;
  static ac_int<2,false> cnt;
  ac_int<2,false> sel;
  if(!cnt[0])//sel for cnt==0 and cnt==2
    sel = 0;
  else if(cnt==3)
    sel = 1;
  else
    sel = 2;
  switch(sel){
  case 0:
    dec2<0>(x,h,y0_int);//read x every 4 clocks with II=2
    break;
  case 1:
    dec2<1>(y0_int,h,y);
    break;
  default:
    break;
  }
  cnt++;
}

Design constraints:
All IO mapped to wire enable interfaces
All arrays mapped to registers
All loops fully unrolled
Top-level design pipelined with II=2
```

The details of example 10-19 are:

- Line 5 defines the intermediate variables used to connect the two decimation stages together. It is declared static because the top-level function be exited after it is written and before it is read. Making it static allows the data to persist between function calls.
- Line 6 defines a static internal counter that controls when each of the decimation stages executes.
- Lines 7 through 13 - to simplify the inlining of the decimation functions, the control is pre-calculated outside of the switch statement that selects which decimator is active. The first decimation stage is run for even values of "cnt" and the second stage runs when "cnt==3". Coding the control this way in general gives better area. Alternatively, "cnt" could be used directly as the "switch" selection, which would mean explicitly writing all of the switch cases. If this is done the decimation functions should be made into components to allow course grain sharing.
- Line 14 - the switch statement case is selected based on the "sel" variable which is calculated based on the value of "cnt".
- Lines 15 through 17 - the input "x" is read for "cnt==0" and "cnt==2". Since the design is pipelined with II=2 this is equal to reading every four clock cycles. The first decimation filter runs at this rate. Since it decimates by two, it produces an output every 8 clock cycles.
- Lines 18 through 20 - The second decimation stage runs every time "cnt" reaches three. Pipelining with an II=2 gives a rate of every eight clock cycles. Since the function decimates by two an output is produced every 16 clock cycles.

Note

Because the filters are in mutually exclusive condition branches their resources can be shared.

Note

Pipelining with II>1 allows even more sharing.

Example 10-19 showed that using a course grained count allowed the different decimation stages to be shared by putting them in mutually exclusive branches of a condition. This approach gives reasonably good sharing without sacrificing high-level coding style. If even better area is needed a lower-level approach can be used. By creating a count that explicitly counts every decimation phase, we can manually schedule the filters to get the minimum number of resources. However to do this the decimators must be coded to match the input/output rate. Example 10-19 had and input rate of four and an output rate of 16. Example 10-20 shown below implements a four-tap decimate by two filter class that assumes an input rate of at least two. E.g. one input every two clock cycles. This class is designed so that it reads an input every other time it is called, and produces an output every fourth time it is called. It only requires a single multiplier to implement the filter. Flags have been added to the class to allow it to be synchronized to the flow of data.

Example 10-20. Manual Decimation Class

```
#ifndef _DEC2_H
#define _DEC2_H
#include <ac_fixed.h>
#include “shift_class.h”
template<int W0, int I0, int W1, int I1>
class dec2_i2{
private:
  shift_class<ac_fixed<W0,I0>,4> regs;
  ac_fixed<W0+W1,I0+I1+2> acc;
  ac_int<2,false> cnt;
  bool vld;
  bool go;
public:
  dec2_i2():vld(false), acc(0), go(false), cnt(0){}
  bool exec(ac_fixed<W0,I0> &x,
            ac_fixed<W1,I1> h[4],
            ac_fixed<W0+W1+2,I0+I1+2> &y,
            bool &vld_in,
            bool &vld_out){
    vld = false;
    if(vld_in)
      go = true;
    if(go){
      if(!(cnt&1))//read with rate 2
        regs << x;
      acc += h[cnt + 2 - (cnt[1]<<2)]*regs[cnt+(1>>cnt[1])-cnt[1]];
      if(cnt==3){//write with rate 4
        y = acc;
        acc = 0;
        vld = true;
      }
      cnt++;
      vld_out = vld;
    }
  }
};
#endif
```

The details of Example 10-20 are:

- Line 5 - the class is templatized to set the integer and fractional bits of the data and coefficients.
- Line 9 - the internal accumulator bit width is set based on the data and coefficient bit widths plus the bit growth due to a four-tap filter.
- Line 10 - an internal count is used to determine the filter phase.
- Lines 11 and 12 - the “vld” flag indicates when data is available at the filter output. The “go” flag controls when the filter can begin executing.
- Lines 15 through 19 - the filter class “exec” method is called to run the filter. The “valid_in” and “valid_out” flags allow multiple instances of the filters to be connected and synchronized.
- Lines 21 through 23 - once the first valid data is detected always run the filter every time “exec” is called.
- Lines 24 and 25 - read the filter input for even values of “cnt”.
- Line 26 - compute one tap times coefficient each time “exec” is called. “cnt”, which determines the filter phase, is sued to select the coefficient and tap based on Figure 10-10 on page 244.
- Lines 27 through 31 - the output is written every fourth call to the function. The “vld” flag is set to indicate valid data available.

Using the manual approach, “cnt” is used to count all of the phases, 16 in this example, of the design while pipelining with II=1. The calls to decimation filters can then be controlled by the current count value. This manual scheduling of the design can best be understood by looking at it in a tabular view.

Manual Scheduling of Two Stage Decimator

cnt	read input	filter1 call	filter1 write output	filter2 call	filter2 write output
0000	x	x			
0001					
0010		x			
0011				x	x
0100	x	x			
0101					

0110		x	x		
0111				x	
1000	x	x			
1001					
1010		x			
1011				x	
1100	x	x			
1101					
1110		x	x		
1111				x	

Table 10-1. The output rate of the design is every 16 clock cycles, so there are 16 clock phases to this design. This allows the first decimator to be called every other clock, and the second decimator to be called every fourth clock. Every fourth call to the decimators in this example produces an output. Table shows that by using the phase count, the decimators can be forced to run in different clock cycles, allowing their resources to be explicitly shared.

Example 10-21 uses the decimation class and implements the schedule shown in Table.

Example 10-21. Low-level Multi-stage Decimation

```
#include <ac_Channel.h>
#include “dec2_i2.hpp”
void dec2_2stage(ac_channel<ac_fixed<8,1> >  &x,
                 ac_fixed<8,1> h[4],
                 ac_channel<ac_fixed<28,7> > &y){
  static  ac_fixed<18,4>  y0_int;
  ac_fixed<28,7> y1_int;
  ac_fixed<8,1>  x_int;
  static ac_int<4,false> phase_cnt;
  static ac_int<2,0> sel_phase;
  static dec2_i2<8,1,8,1> f0;
  static dec2_i2<18,4,8,1> f1;
  static bool f0_vld_in, f0_vld_out, f1_vld_out;

  if(!phase_cnt[0])//if even counts
    sel_phase = 0;
  else if(phase_cnt.slc<2>(0)==3)//if every 4th odd count
    sel_phase = 1;
  else
    sel_phase = 3;//do nothing
  if(phase_cnt.slc<2>(0)==0){//read a rate of 4
    x_int = x.read();
```

```
23      f0_vld_in = true;
24    }else
25      f0_vld_in = false;
26    switch(sel_phase){
27    case 0:
28      f0.exec(x_int,h,y0_int,f0_vld_in,f0_vld_out);
29      break;
30    case 1:
31      f1.exec(y0_int,h,y1_int,f0_vld_out, f1_vld_out);
32      if(f1_vld_out)
33        y.write(y1_int);
34      break;
35    default:
36      break;
37    }
38     phase_cnt++;
39  }
```

```
Design constraints:
All IO mapped to wire enable interfaces
All arrays mapped to registers
All loops fully unrolled
Top-level design pipelined with II=1
```

The details of Example 10-21 are:

- Line 9 - a four bit counter is used to count all sixteen phases.
- Lines 11 and 12 - two static instances of the decimation filter class from Example 10-20 are used.
- Lines 15 through 20 - the phase selection control for the design is computed here instead of passing "phase_cnt" directly to the switch statement. This reduces the number of cases for the switch statement, and should give better area in general.
- Lines 21 through 25 - the input must be read every fourth call to the top level function. The lower two bits of "phase_cnt" are masked to see if the input should be read. When the input is read the "vld_in" flag that connects to the first decimator is set equal to true, synchronizing the flow of data.
- Lines 26 through 37 - the two decimation filters are called based on "sel_phase" which is set based on "phase_cnt" and Table. The final output is only written when the "vld_out" flag is set by "f1".

Chapter 11

FFT Transform

Introduction

The Fast Fourier Transform (FFT) is one of the best algorithms to illustrate why high-level C++ synthesis is simply an evolution in hardware design, still requiring the expertise of an RTL or system designer, while facilitating rapid implementation of complex memory architectures and control. It is also the perfect example to help understand where many of the common misconceptions about HLS come from.

Because the FFT is a well understood algorithm by both algorithm developers and RTL designers, it is more often than not the first design chosen to evaluate when learning HLS. This highly ambitious, yet sometimes flawed, approach often consists of taking a purely algorithmic description of a floating point FFT and synthesizing it to RTL, with the expectation being an optimal hardware implementation.

Most HLS tools are capable of producing a working design from this pure algorithmic FFT description, which is usually either too slow, too big, or both. Because the source language description contains no explicit memory architecture, it becomes the job of the HLS tool to optimize the design solely based on user-applied constraints such as array to memory mapping, memory splitting or interleaving, loop unrolling, and pipelining. The non-linear array index access pattern of the pure algorithmic description prevents many of these constraint-based optimizations. Furthermore, without the use of bit-accurate data types, the area is undoubtedly larger that what one would expect from a hand crafted RTL design. Because this issue is so important, it's worth repeating that the implementation/architectural details required in the source language description parallels what RTL designers embed in their designs today. The difference is that the abstractness of C++ takes care of the low-level details automatically, allowing designers to focus more on the architecture and control. The RTL designer's expertise is an essential ingredient in achieving good quality of results using HLS.

The pure algorithmic description of the FFT, unlike the FIR filter, is not a good algorithm for performing architectural exploration. This is primarily due to the non-linear indexing used by the algorithm. While in theory it is possible to partially explore a register-based FFT algorithm, trading off area versus performance, it usually results in unacceptably large area. Memory-based algorithms, which cannot be easily explored, suffer from sub-optimal performance because of memory access bottle-necks created by non-linear indexing.

There is no “one size fits all” when it comes to FFTs. The underlying architecture depends on many factors from the use of registers v.s. memories to the required throughput and area. It is beyond the scope of this chapter to cover all of these implementations. The goal of this chapter is to present an efficient memory-based radix-2 in-place FFT that illustrates the types of analysis and C++ code transformations that are required to generate good quality hardware. These techniques can be applied to many other types of FFTs.

Radix-2 FFT

An FFT is an algorithm for computing the discrete Fourier transform (DFT) that is more efficient than the straight forward computation of the DFT [1]. The DFT is computed as:

$$X(k) = \sum_{n=0}^{N-1} x(n) W_N^{nk}$$

Where $W_N = e^{-j2\pi/N}$. The DFT requires on the order of N^2 operations. By recursively splitting the FFT computation into odd and even parts the number of operations can be reduced to the order of N*LOG2(N). The summation then becomes:

$$X(k) = \sum_{r=0}^{N/2-1} x(2r) W_{N/2}^{rk} + W_N^{k} \sum_{r=0}^{N/2-1} x(2r+1) W_{N/2}^{rk}$$

$$= X_{even}(k) + W_N^{k} X_{odd}(k)$$

where the first summation is an N/2 point DFT over the even indexed inputs and the second summation is an N/2 point DFT over the odd indexed inputs. This procedure is applied recursively and leads to a computation requiring stages as shown in Figure 11-1. For example the outputs X(1) and X(5) can be derived as

$$X(1) = X_{even}(1) + W_8^{1} X_{odd}(1) = X_{even}(1) + W_8^{1} X_{odd}(1)$$

$$X(5) = X_{even}(5) + W_8^{5} X_{odd}(5) = X_{even}(1) - W_8^{1} X_{odd}(1)$$

The above two computations correspond to the computation in the last stage of Figure 11-1 that takes input 1 from both the upper (even) and lower (odd) four point outputs of stage 2. The complex multiplication by W_8^{1}, called twiddle factor, is common for both expressions leaving an addition and a subtraction which together are called a radix-2 butterfly [1].

Figure 11-1. Radix-2 FFT Data Flow Diagram

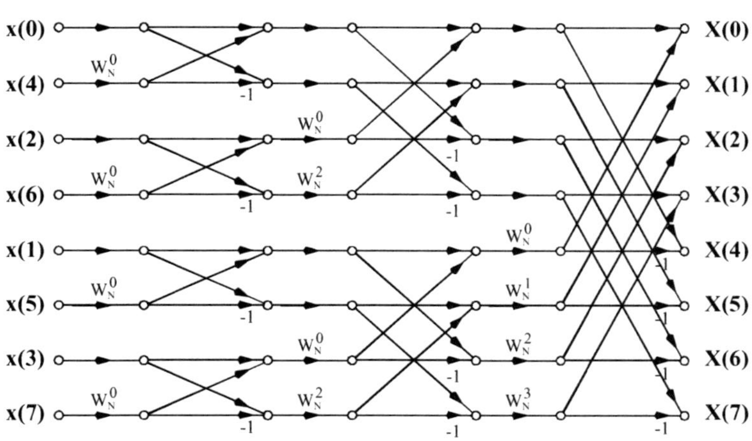

Floating Point Radix-2 In-place FFT

One of the most commonly used algorithmic implementations of the radix-2 FFT is a floating-point, in-place implementation. The term "in-place" means that the algorithm uses a single array/memory to compute each stage of FFT shown in Figure 11-1. The C++ implementation can be expressed very compactly when written using a purely algorithmic style. Example 11-1 shows a typical floating point implementation. There are a number of coding style problems with this implementation that make it unsuitable for synthesis. The details of Example 11-1 are:

1. Lines 5,6, and 8 - the algorithm uses double precision. In other words it has not been quantized.
2. Lines 24 and 25 - the cos and sin functions from <math.h> are used to compute the twiddles. These are not synthesizable. Additionally it is more likely that a hardware implementation uses a lookup table to store the needed twiddle factors.
3. Lines 26 and 27 - the complex multiply of the butterfly uses four multipliers. This is inefficient since it can be done using three multipliers.
4. Entire design - the complex arithmetic has been split into real and imaginary parts. This leads to more lines of code.

Example 11-1. Floating-point Radix-2 FFT

```
1   #include "fft_float.h"
2   #pragma design top
3   void fft(
4            double x_r[FFT_SIZE],
5            double x_i[FFT_SIZE])
6   {
7     double t_r, t_i,cos_twid,sin_twid;
8     int n1;
9     int idx;
10    int n2 = FFT_SIZE/2;
11
12    n1 = 0;
13    n2 = 1;
14    idx = FFT_SIZE;;
15    for(int i=0;i< FFT_STAGES;i++){
16      n1 = n2;
17      n2 = n2 + n2;
18      idx>>=1;
19      for(int j=0;j< FFT_SIZE/2;j++){
20        int k=j;
21        for(int kk=0;kk<FFT_SIZE/2;kk++){
22          cos_twid = cos(2*pi*j*idx/FFT_SIZE);
```

```
                sin_twid = sin(2*pi*j*idx/FFT_SIZE);
                t_r = cos_twid * x_r[k+n1] + sin_twid * x_i[k+n1];
                t_i = cos_twid * x_i[k+n1] - sin_twid * x_r[k+n1];
                x_r[k+n1] = x_r[k] - t_r;
                x_i[k+n1] = x_i[k] - t_i;
                x_r[k] = x_r[k] + t_r;
                x_i[k] = x_i[k] + t_i;
                k+=n2;
                if(k>=FFT_SIZE-1) break;
            }
            if(j==n1-1) break;
        }
    }
}
```

Aside from the coding style issues, there is a more fundamental architectural problem with the implementation of Example 11-1 when the storage arrays "x_r" and "x_i" are mapped to singleport memories. Ideally a radix-2 memory mapped in-place FFT should be able to perform one butterfly per clock cycle for optimal throughput. However this is not possible because there is a memory address conflict that occurs for butterfly reads and writes [2]. To put it simply each butterfly is trying to read/write the singleport memory twice in the same clock cycle (See "Memories" on page 104). Using a true dual-port RAM would allow this design to be pipelined with II=1, but the area cost would be prohibitive. Figure 11-2 shows the DFG for an 8-point radix-2 FFT where the array data "a0, a1, ... a7" are shown for the different butterfly stages. Mapping the array to memory results in conflicting addresses for the memory as the butterfly attempts to simultaneously read/write the memory. The solution is use two singleport memories to implement the FFT. However this requires reordering of the input data as well as restructuring the data flow graph so that there is only one read/write to the memories for each butterfly computation.

Figure 11-2. Radix-2 FFT Data Flow Graph

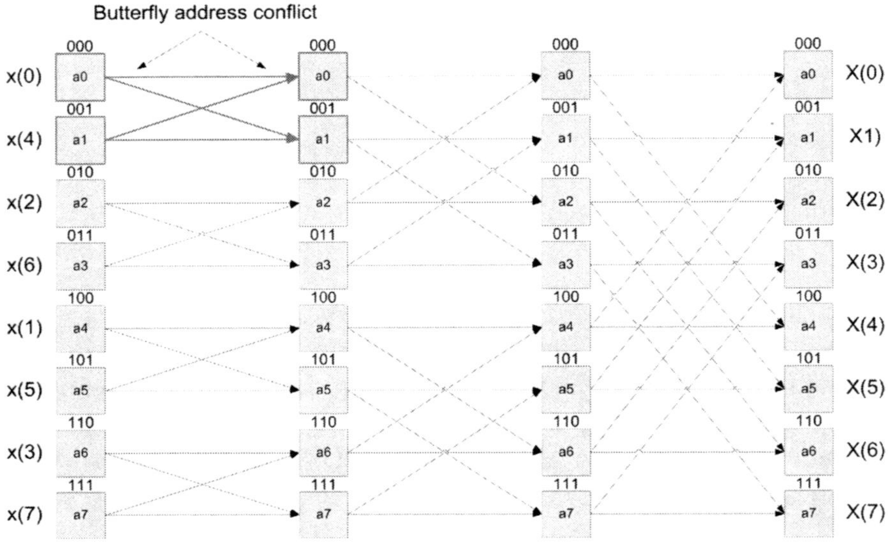

Figure 11-3 shows the radix-2 FFT data flow graph after it has be restructured to avoid any address conflicts. The restructuring of the DFG, along with reordering the input data, allows two singleport RAMs to be used so that the butterfly can execute every clock cycle.

Figure 11-3. Re - structured DFG for Radix-2 FFT

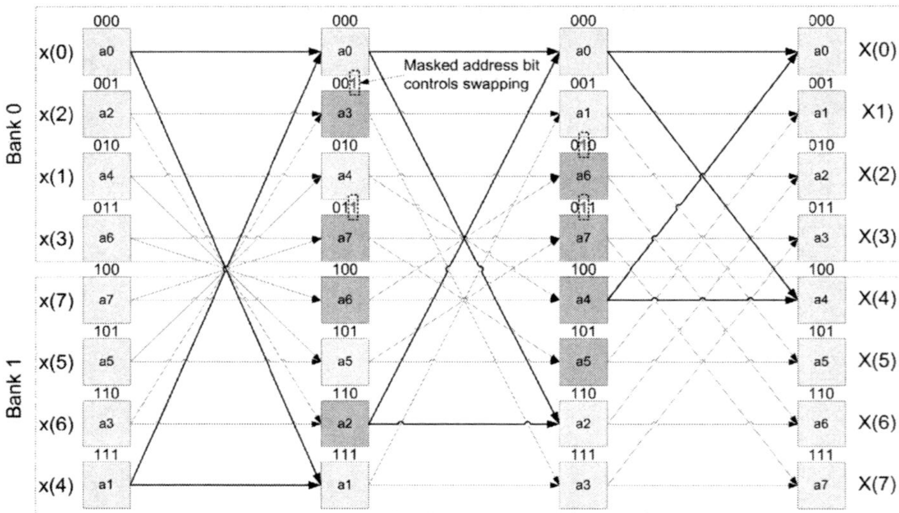

The bit-reversed input data in the first butterfly stage is reordered into even and odd halves. By mapping the even and odd halves to separate singleport RAMs, the first stage butterflies can run every clock cycle. The address generated to index the Bank0 memory is shifted and masked for writes and reads during each stage. If the masked address bit, shown in Figure 11-2, is set equal to one, the butterfly write/read data is "swapped". By the time the last stage is reached the data has been written out in the original order.

Example 11-2. Fixed-point Radix-2 FFT

```
#include "fft_fixed.h"
void fft(
         ac_complex<dType > x_l[FFT_SIZE/2],
         ac_complex<dType > x_u[FFT_SIZE/2])
{
  ac_complex<dType > x_tmp,data_tmp, data_l, data_u;
  ac_complex<mType > t;
  ac_int<FFT_STAGES+1,false> n1, n2;
  ac_int<FFT_STAGES+1,false> idx;
  ac_int<FFT_STAGES,false> addr_mask = 1;
  ac_int<FFT_STAGES,0> idx_l=0;
  ac_int<FFT_STAGES-1,0> idx_u = 0;
  n1 = 0;
  n2 = 1;
  idx = FFT_SIZE;
  for(int i=0;i< FFT_STAGES;i++){//stage
    idx_l = 0;
    n1 = n2;
    n2 = n2 + n2;
    idx>>=1;
    for(int j=0;j< FFT_SIZE/2;j++){//segment
      int k=j;
      for(int kk=0;kk<FFT_SIZE/2;kk++){//butterfly
        idx_u = idx_l^(-1<<i);
        data_l = x_l[idx_l];
        data_u = x_u[idx_u];
        swap(addr_mask>>1, idx_l, data_l, data_u);
        t = complex_mult(twiddle(j * idx) ,data_u);
        x_tmp = data_l;
        data_u = scale(x_tmp - t);
        data_l = scale(x_tmp + t);
        swap(addr_mask, idx_l, data_l, data_u);
        x_u[idx_u] = data_u;
        x_l[idx_l] = data_l;
        k+=n2;
        idx_l += (1<<i);
        if(idx_l[FFT_STAGES-1]){//if idx overflows, wrap
          idx_l[FFT_STAGES-1] = 0;
          idx_l += 1;
        }
        if(k>=FFT_SIZE-1) break;
      }
```

```
43        if(j==n1-1) break;
44      }
45      addr_mask <<= 1;
46    }
47  }
```

Example 11-2 shows a fixed-point radix-2 FFT that implements the memory architecture outlined in Figure 11-3. The details are:

1. Lines 3 and 4 define the two memories required by the architecture. The real and imaginary parts are combined using the Algorithmic C complex data type "ac_complex". This type is templatized to allow user specification of the base data type, which in this example is type defined as "dType" in a global typedefs header file. Doing this allows one to switch between float and fixed point types for debugging. The typedefs are shown below in Example 11-3 on page 268.
2. Lines 8 through 12 define a number of index, mask, and counter variables that depend on the number of stages of the FFT. The number of FFT stages is determined by taking log2 of the FFT size. This is done in a header file (Example 11-4 on page 268), using the log2_ceil function supported by the Algorithmic C data types.
3. Lines 18 through 20 are the same as the original floating point FFT. They control the segment and butterfly loop iterations as well as the index into the twiddle tables.
4. Line 24 - the index into the upper, or Bank1, memory is derived from the index into the lower memory. It counts in the opposite direction as the lower memory index, and the starting position is controlled by the FFT stage of (-1<<i). This expression is derived from analyzing the DFG of Figure 11-3.
5. Lines 25 and 26 read a complex value from both the lower and upper memories. This data is used for the butterfly computation. Note that these memories are only read once per butterfly loop iteration.
6. Line 27 calls the "swap" function (Example 11-5 on page 268) which masks the read address for the lower memory. If the masked bit is set the data is swapped as shown in Figure 11-3.
7. Line 28 performs the complex multiplication of the complex data against the complex twiddle. The twiddles are read from a constant

array, Example 11-6 on page 269, and the complex multiply is implemented with minimum resources, Example 11-9 on page 271.

8. Lines 30 and 31 call the "scale" function, Example 11-10 on page 271, which scales the data computed for each butterfly stage by dividing by two. This is done to avoid overflow.
9. Line 32 masks the write address for the lower memory. If the masked bit is set, the lower memory and upper memory data is swapped before writing.
10. Lines 33 and 34 write the lower and upper memories. Each memory is only written once per butterfly loop iteration.
11. Lines 36 through 40 - the lower memory index is computed based on the current FFT stage. The upper bit of the index is checked to detect when the count overflows and the count is "wrapped" around to the begging of the memory.

Example 11-3. FFT Types

```
#ifndef __FFT_TYPES__
#define __FFT_TYPES__
#include <ac_fixed.h>
#include <ac_complex.h>

#define FIXED
#ifdef FIXED
typedef ac_fixed<14,2> dType;
typedef ac_fixed<29,5> mType;
typedef ac_fixed<15,2> tType;
#else
typedef double dType;
typedef double bType;
typedef double mType;
typedef double tType;
#endif
#endif
```

Example 11-3 shows how a global typedef file can be used to easily switch between fixed point and floating point. This is often useful when debugging a design.

Example 11-4. FFT Constant Header File

```
#ifndef __FFT_CONSTS__
#define __FFT_CONSTS__
#include <ac_int.h>
const int FFT_SIZE = 1024;
const int FFT_STAGES = ac::log2_ceil<FFT_SIZE>::val;
const int MASK_BITS = FFT_STAGES-1;
```

```
const double pi = 3.1415;
#endif
```

Example 11-4 shows how the design can be parameterized based on the size of the FFT. The use of the built-in helper function form the Algorithmic C data types allow static computation of the log2(FFT_SIZE).

Example 11-5. Data Swapping Based on Address Mask

```
#include "fft_fixed.h"
void swap(int addr_mask, int idx_l, ac_complex<dType > &data_l,
          ac_complex<dType > &data_u){
  ac_complex<dType > data_l_tmp;
  bool swap_d;
  data_l_tmp = data_l;
  swap_d = (addr_mask & idx_l);
  data_l = swap_d ? data_u : data_l;
  data_u = swap_d ? data_l_tmp : data_u;
}
```

Example 11-5 simply "and"s the incoming address against the mask and swaps the lower and upper data if the masked bit is set.

Example 11-6. Computing the Twiddles

```
#include "fft_fixed.h"
ac_complex<tType> twiddle(int n){
  ac_complex<tType> tmp;

  tmp.r() = cos_lookup(n);
  tmp.i() = -sin_lookup(n);
  return tmp;
}
```

Example 11-6 computes the sin and cos, or twiddles, for the FFT. It does this by simply reading them from a lookup table, which is implemented as a constant array. Because the quadrants of the sin and cos are even and odd symmetrical it is only necessary to store one quarter of the waveforms. The table lookup functions are shown below.

Example 11-7. Computing the Sin Twiddles

```
#include "fft_fixed.h"
tType sin_lookup(int n){
tType sin_table[FFT_SIZE/4+1] = {
  #include "sin_qtable.txt"
};
 tType tmp;
```

```
7     int idx;
8     bool sign;
9     if(n<=FFT_SIZE/4){
10      idx = n;
11      sign = 0;
12    }
13    else if( n<FFT_SIZE/2){
14      idx = FFT_SIZE/4-n%(FFT_SIZE/4);
15      sign = 0;
16    }
17    else if(n<3*FFT_SIZE/4){
18      idx =  n%(FFT_SIZE/4);
19      sign = 1;
20    }
21    else{
22      idx = FFT_SIZE/4-n%(FFT_SIZE/4);
23      sign = 1;
24     }
25     return sign ? (tType)-sin_table[idx] : (tType)sin_table[idx];
26  }
```

The details of Example 11-7 are:

1. Lines 3 through 5 show the constant array that stores one quarter, plus one sample, of the sin waveform. The lookup table uses the technique covered in "Lookup Tables (LUT)" on page 155.
2. Lines 9 through 12 implement the sin form 0 to pi/2.
3. Lines 13 through 16 compute the sine for pi/2 to pi. Note that the index is reversed.
4. Lines 17 through 20 compute the sin for pi to 3/2pi. In this case "sign" is set so that the sign of the result is made negative.
5. Lines 21 through 24 compute the sin for 3/2pi to 2*pi. The index is reversed and the "sign" bit is also set to make the result negative.
6. Line 25 returns either "sin[idx]" or "-sin[idx]" depending on the "sign" bit.

Example 11-8 computes the cos using the same technique as Example 11-7.

Example 11-8. Computing the Cos Twiddles

```
#include "fft_fixed.h"
tType cos_lookup(int n){
tType cos_table[FFT_SIZE/4+1] = {
  #include "cos_qtable.txt"
};
 tType tmp;
```

```
  int idx;
  bool sign;
  if(n<=FFT_SIZE/4){
    idx = n;
    sign = 0;
  }
  else if(n<FFT_SIZE/2){
    idx = FFT_SIZE/4-n%(FFT_SIZE/4);
    sign = 1;
  }
  else if(n<3*FFT_SIZE/4){
    idx =  n%(FFT_SIZE/4);
    sign = 1;
  }
  else{
    idx = FFT_SIZE/4-n%(FFT_SIZE/4);
    sign = 0;
  }
  return sign ? (tType)-cos_table[idx] : (tType)cos_table[idx];
}
```

Example 11-9 implements the complex multiply using a more efficient method that reduces the number of multiplications to three[3].

Example 11-9. Complex Multiply

```
#include "fft_fixed.h"
#pragma map_to_operator
ac_complex<mType > complex_mult(ac_complex<tType > twiddle, ac_complex<dType > data){
  tType a,b;
  dType c,d;
  mType e;
  ac_complex<mType > tmp;
  a = twiddle.r();
  b = twiddle.i();
  c = data.r();
  d = data.i();

  e = a*(c-d);
  tmp.r() = d*(a-b) + e;
  tmp.i() = c*(a+b) - e;
  return tmp;
}
```

Example 11-10 implements the scaling function that is used to scale the data from each butterfly. This is a very "crude" method for preventing overflow. More advanced methods are beyond the scope of this chapter.

Example 11-10. Scaling

```
#include "fft_fixed.h"
ac_complex<dType > scale(ac_complex<mType > data){
```

```
    ac_complex<mType > tmp;

    tmp = data;
    tmp.r() = tmp.r()>>1;
    tmp.i() = tmp.i()>>1;
    return tmp;
}
```

Some Final Thoughts

The radix-2 FFT presented in this chapter could be considered a relatively low performance design since it is only capable of executing one butterfly per clock cycle. Higher performance FFTs require different architectures than what was covered. However the methods used here to go from a floating point algorithm to a fixed point implementation are applicable to most designs. The important point to take away from this chapter, as well as the book, is to understand that a pure algorithmic description is not suitable for creating good quality hardware. The architectural and control details must be part of the C++ implementation to achieve results that are comparable to hand-coded RTL. This means approaching the design creation using many of the same skills that designers use today. No matter what the algorithm, a designer using HLS needs to be asking the question "What would the hardware look like if I did this by hand"? Understanding the block level structure of what the hardware should look like is usually sufficient to know how the C++ should be organized. This means capturing the memory architecture in the C++ code.

Remember, High Level Synthesis is only a hardware design methodology. It requires a "hardware designer" to realize the productivity gains over RTL design. It isn't going to turn everyone into hardware designers, but it will allow hardware designers to match the ever-increasing complexity of ASIC design.

References

1. Andres Takach. Creating C++ IP for High Performance Hardware Implementations of FFTs. DesignCon2002.
2. L. G. Johnson. Conflict Free Memory Addressing for Dedicated FFT Hardware. IEEE TRANSACTIONS ON CIRCUITS AND SYSTEMS-11: ANALOG AND DIGITAL SIGNAL PROCESSING, VOL. 39, NO. 5, MAY 1992.

3. Preston A. Jackson, Cy P. Chan, Jonathan E. Scalera, Charles M. Rader, and M. Michael Vai - A Systolic FFT Architecture for Real Time FPGA Systems. MIT Lincoln Laboratory 244 Wood ST, Lexington, MA 02420

Lightning Source UK Ltd.
Milton Keynes UK
22 July 2010

157351UK00001B/1/P

9 781450 097246